Fast Evolution

Adaptation on Human Timescales

Team Trex

Published by NobleTrex Press

For permissions and other inquiries, write to:

P.O. Box 3132, Framingham, MA 01701, USA

Contents

1 Evolution in Overdrive **9**

1.1 The Human Asteroid 9

1.2 Darwin on Fast Forward 16

1.3 The Illusion of Change 24

2 The Shape-Shifter's Toolkit **35**

2.1 The Warehouse of Variance 35

2.2 The Founder's Gamble 42

2.3 Highways of Heredity 50

3 Invisible Armies **59**

3.1 The Blueprint Swappers 59

3.2 Fortresses of Slime 66

3.3 The Infinite Arms Race 73

4 Concrete Jungles **83**

4.1 The Furnace of Asphalt 83

4.2 Scaling the Glass Cliffs 93

4.3 Evolution Underground 100

5 Chemical Warfare **107**

5.1 The Moth and the Smokestack 107

5.2 Convergent Concrete Jungles 114

5.3 The Toxic Trade-Off 123

6 The Infinite Feast **133**

6.1 The Chemical Crucible 133

6.2 Fortresses of Corn and Cotton 141

6.3 Outsmarting the Algorithm 148

7 Time Travel in a Flask **157**

7.1 The 75,000-Generation Replay 157

7.2 Reading the Source Code 165

7.3 The Crowd-Sourced Laboratory 173

8 Designing the Future **181**

8.1 The Heavy Price of Survival 181

8.2 Forecasting the Unseen 190

8.3 The Architecture of Resilience 197

Introduction

In the early 1990s, the island of Kauai fell silent. For roughly four million years, the nights on this Hawaiian island had vibrated with the courtship songs of male teleogryllus crickets-a rhythmic, chirping chorus that signaled the tropical evening just as reliably as the setting sun. Then, over the span of just a few years, the noise stopped. The crickets were still there, hopping through the grass in teeming numbers, but they were opening and closing their wings without making a sound.

To a casual observer, it might have seemed like a sudden onset of collective shyness. To evolutionary biologist Marlene Zuk, who had been studying these populations, it was a baffling mystery. The investigation revealed a grim driver behind the silence: a parasitic fly, *Ormia ochracea*, which had been introduced to the islands. This fly hunted by sound. It homed in on the cricket's chirp, landed on the singer's back, and deposited larvae that would burrow inside and devour the host from within.

Singing, for the crickets, had become a death sentence.

In a timeline that would have made Charles Darwin drop his teacup, the cricket population responded. A random genetic mutation appeared that flattened the wings of male crickets, erasing the rigid structures used to make noise. Silent males couldn't attract females as easily-they had to rely on clustering near the few remaining singers to intercept mates-but they didn't get eaten. The selection pressure was so intense that in fewer than twenty generations, the silent mutation swept through the population. The crickets had not just changed their behavior; they had physically re-engineered their bodies to survive a new reality.

We are taught in school that evolution is a saga of deep time. We learn about it through the fossil record, viewing it as a majestic, glacial process where dinosaurs slowly transform into birds over tens of millions of years. We think of it as history. But the story of the Kauai crickets tells us that evolution is not just history; it is current events. It is happening right now, in the weeds of your backyard, in the bacteria on your smartphone screen, and in the pigeons cooing on the ledge of a skyscraper.

This book is about that speed. It is about the realization that when the environment changes abruptly, life doesn't always wait millions of years to catch up. And nothing changes the environment quite like us.

We are living in a unique biological moment. Humans have become the world's dominant evolutionary force. We act as a strange, multifaceted asteroid that hasn't just hit the planet once but continues to reshape it every day. We build concrete archipelagos called cities; we saturate the soil with nitrogen and poisons; we transport species across oceans in the ballast of ships; and we flood the atmosphere with carbon. In doing so, we have inadvertently cranked the engine of natural selection up to overdrive.

For a long time, there was a scientific assumption that because human environments are "artificial," the biology happening within them is somehow less valid-that the raccoon raiding a dumpster is less noble, biologically speaking, than the wolf hunting elk. But nature makes no such distinction. DNA does not care if a selection pressure comes from a shifting glacier or a new brand of pesticide. It only knows survival.

What we are witnessing today is a global experiment in rapid adaptation. The creatures and plants that share our world are engaging in a desperate, creative form of biological improvisation. They are rewriting their genetic codes to survive the world we have built. This phenomenon is often called "contemporary evolution" or "rapid evolution," and it challenges the comfortable distance we usually place between ourselves and the natural world. We are not just observers of nature; we are its most aggressive editors.

Consider the layout of a modern city. To us, a city is a place of commerce and culture, a grid of transit and housing. To an evolutionary biologist, a city is a landscape of extreme selection pressures. It is a canyon system of glass and steel where the temperature is hotter, the light never truly fades, and the food sources are vastly different from the wild.

In response, the biology of the urban world is diverging from its rural counterparts. In Puerto Rico, Shane Campbell-Staton and his colleagues have found that city-dwelling anole lizards are evolving different bodies than their forest brethren. The city lizards possess longer limbs and stickier toe pads-specialized equipment for gripping smooth concrete and painted metal rather than rough tree bark. They have become urban mountaineers, adapted physically to the architecture of human civilization.

This isn't just about morphology; it's about physiology and behavior. Across Europe, urban blackbirds are singing at higher pitches to cut through the low-frequency rumble of traffic, and they have altered their migration genes to become sedentary residents of our heated metropolises. In the tunnels beneath London, a distinct subterranean form of mosquito has emerged-the so-called "London Underground mosquito." Separated from the surface world since the construction of the Tube, these insects have evolved to feed on rats and commuters instead of birds, and they can no longer successfully breed with their above-ground ancestors. In the blink of an evolutionary eye, human infrastructure has created the conditions for speciation.

This rapid reshaping of life is fascinating, but it is also a warning. When we force evolution to move this fast, the results are not always to our liking. We are currently locked in a high-stakes arms race with the microscopic world. For decades, we believed that antibiotics were a permanent victory over infectious disease. We forgot that bacteria are the original masters of rapid evolution.

Inside hospitals and livestock facilities, we have created the perfect training grounds for resistance. By flooding these environments with lethal drugs, we have systematically killed off the weak and left only the strong-the mutants capable of pumping out the poison or shielding themselves behind impenetrable biofilms. As we will see when we examine the "invisible armies" of the microbial world, bacteria are not solitary actors. They are biological traders, swapping genes like Pokémon cards. A harmless soil bacterium can pass a resistance gene to a deadly pathogen on a handshake. The result is the rise of "superbugs" that threaten to return medicine to the dark ages. This is evolution looking us in the eye and demanding a response.

The same dynamic plays out in our agricultural heartlands. We spray herbicides to clear our fields, expecting a sterile canvas for our crops. Instead, we have bred "superweeds"-plants like the Palmer amaranth, which can grow inches in a day and shatter the steel of combine harvesters. These plants have evolved metabolic shields to neutralize our chemicals, forcing farmers into an escalating cycle of toxicity and resistance. We try to engineer the genome of our crops to be pest-proof, only to watch the bollworm and the rootworm evolve workarounds in a matter of seasons. It is a stark lesson in the concept that there is no such thing as a "solution" in biology, only a temporary strategy.

To understand how this happens, we have to look under the hood of life itself. We need to explore the "shape-shifter's toolkit"-the mechanisms of mutation, recombination, and gene flow that provide the raw material for these transformations. It turns out that life is far more messy and modular than the clean diagrams in high school textbooks suggest. Genetic variation is everywhere, waiting for the right moment to shine. Sometimes, the genes necessary to survive a new poison are already floating in the population at low levels, just waiting for the environment to change. Other times, species rely on the "migrant gene," where a traveler from a distant population brings a crucial adaptation that saves the locals from extinction.

What makes this era of science so thrilling is that we no longer have to guess at these processes. We can watch them happen. We have entered the age of "time travel in a flask." In laboratories like that of Richard Lenski at Michigan State University, researchers have been tracking the evolution of *E. coli* bacteria for over 75,000 generations. They have frozen the "ancestors" in time, allowing them to compete the great-great-grandchildren against their own forefathers to see exactly how much fitter they have become. We can sequence the DNA of a virus in real-time as it spreads across the globe, watching the letters of its genetic code shift and change as it jumps from host to host. We are no longer reconstructing the crime scene of evolution; we are watching the security footage.

However, recognizing the speed of evolution requires a shift in our own perspective. We often suffer from "shifting baseline syndrome," accepting the state of the natural world as we found it in our youth as "normal." We fail to see that the fish are getting smaller because our nets catch the big ones, leaving the genes for smallness to propagate. We miss the fact that the tuskless elephants of Mozambique are becoming more common because poachers have systematically removed the genes for tusks from the gene pool. We look at a field of white clover and see a static lawn, unaware that the plants in the city center have evolved to stop producing cyanide-a defense mechanism against herbivores that is useless in the warmth of the city where snow doesn't insulate them from frost.

Why does this matter to you? Because understanding fast evolution is the key to our own future. If we continue to view the natural world as a static painting that we can deface without consequence, we will continue to be blindsided by the backlash. We will be outwitted by viruses, overrun by resistant pests, and we will watch the biodiversity that sustains us collapse because it cannot keep up with the pace of change we inflict.

But if we accept that we are driving the bus of evolution, we can learn to steer. We can design our hospitals to discourage resistance rather than breed it. We can plan our cities to be corridors of connectivity rather than genetic dead-ends. We can develop agricultural practices that use evolutionary principles to manage pests rather than trying to annihilate them. This is the promise of "evolutionary forecasting"- the audacious idea that we can predict the path of adaptation and prepare for it.

The journey ahead will take us from the smog-choked history of the Industrial Revolution, where the peppered moth first taught us the power of pollution, to the radioactive forests of Chernobyl, where tree frogs have turned pitch black to shield themselves from radiation. We will look at the genetic distinctiveness of New York City rats and the strange, toxic tolerance of fish in the Hudson River.

We are living in an era where the boundary between the "natural" and the "artificial" has dissolved. The concrete jungle is just as much an ecosystem as the Amazon rainforest; it just operates under different rules-rules that we wrote, and rules that nature is learning to exploit.

So, forget the idea of evolution as a slow, dusty museum exhibit. Life is fast, fluid, and fiercely competitive. It is happening in the cracks of the sidewalk and the sterile air of the intensive care unit. The planet is adapting to the human age. The question is: can we adapt to what it becomes?

Chapter 1

Evolution in Overdrive

We often imagine evolution as a majestic, glacial process visible only in the fossil record, but the biological world is currently rewriting its genetic code at breakneck speed. By transforming the planet, humanity has inadvertently hit the fast-forward button on natural selection, forcing life to adapt or perish within the span of a single human lifetime.

1.1 The Human Asteroid

Imagine an asteroid not as a single flash of catastrophe, but as a continuous presence: a force that rearranges landscapes, chemistry, and survival rules faster than most species can respond. That is the evolutionary role humans now play. The comparison is uncomfortable, yet clarifying. A meteor ends its work in an afternoon; human activity keeps striking the biosphere every day, in millions of places, through heat, toxins, harvest, fences, antibiotics, plastics, light, noise, and a thousand kinds of habitat rearrangement. Extinction is the most visible outcome, but it is not the only one. Where populations persist, the survivors are often those whose inherited traits happen to fit the new human-made world. Natural selection becomes less like a slow sculptor and more like triage.

The deep past contains plenty of fast environmental change, but much of it was either rare (a major impact, a sudden volcanic episode) or spatially patchy (a local drought, a shifting river course). Human change is different in three crucial ways. First, it is *globally synchronized*: similar pressures appear on different continents at roughly the same time, from pesticide exposure in fields to artificial light at night in cities. Second, it is *novel*: no previous epoch offered chlorinated drinking water, leaded gasoline residues, neonicotinoids, or a worldwide antibiotic haze in hospitals and farms. Third, it is *repeated and directional*: the same stressor hits again and again, every season, every prescription, every trawl, every commute. Selection becomes not only intense, but dependable.

This intensity is easiest to see where humans create what might be called evolutionary *kill zones*: places where most individuals die or fail to reproduce unless they carry just the right combination of traits. A kill zone can be literal, like a field sprayed with herbicide, or statistical, like a fishery that removes the largest adults year after year, leaving the rest to breed. It can also be invisible to us, like a hospital ward where bacteria encounter a shifting cocktail of antibiotics and disinfectants. The common feature is that survival is no longer mainly about avoiding predators, finding food, or enduring winter. It becomes about clearing a new obstacle course built at human speed.

A useful way to picture this is as a set of sieves that act on a cloud of variation. Real populations are not uniform; they are messy distributions of different sizes, behaviours, immune systems, metabolisms, and timing. Humans push that cloud through filters so narrow that only a particular sliver of variation makes it out the other side. The survivors then form the next generation, and what began as a minority trait can become ordinary with startling speed.

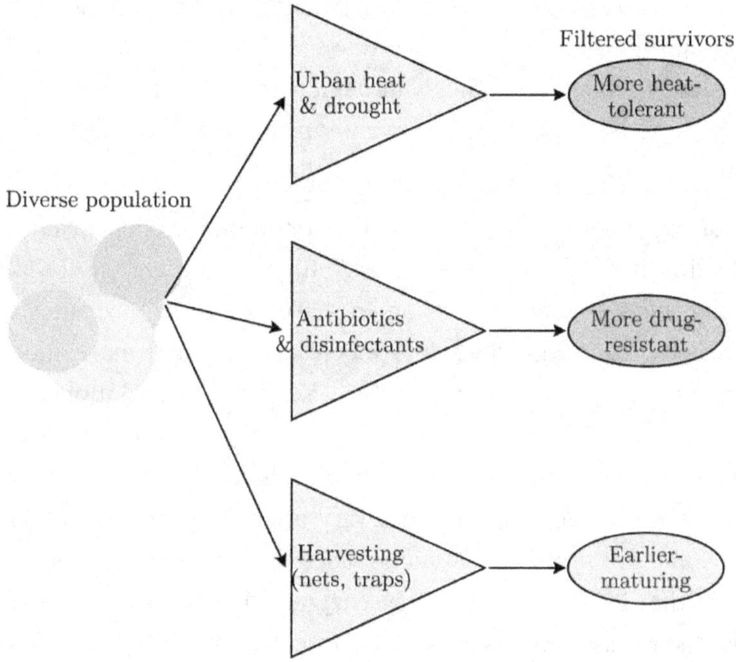

Human activity often acts like a set of strong filters. Each filter removes most of the original variation and leaves a different, narrower set of survivors to reproduce.

Once you start looking for kill zones, they appear everywhere. Industrial agriculture is one. A modern crop field is a simplified habitat: vast acreage of a single plant genotype or a small handful of genotypes, planted at the same time, fertilized to remove nutrient limitation, irrigated to reduce drought, and protected by chemical shields. That shield does not simply remove pests; it selects for pests that can slip through. The outcome is so predictable that it has become a grim rhythm of innovation: a new pesticide works beautifully, then within years or decades resistance spreads, doses rise, mixtures are tried, and the arms race continues. The same logic holds for herbicides and weeds. A field treated with a single herbicide is an ecological desert for susceptible weeds, but a paradise for any variant that can tolerate the chemical. Many of those variants already exist at low frequency, or can arise quickly through mutation. When nearly every competitor is removed, the resistant type is not merely surviving; it is inheriting an empty stage.

Medicine is another kill zone, and a more intimate one, because the selection happens inside bodies. Antibiotics are among the most transformative technologies humans have ever developed: they convert once-lethal infections into inconveniences and make modern surgery and chemotherapy possible. Yet from an evolutionary perspective, an antibiotic is not primarily a cure. It is a filter. The drug does not aim at health in the abstract; it aims at a molecule, a pathway, a ribosome, a cell wall. Bacteria vary in how vulnerable that target is. Some are protected by membranes that block entry; some can pump the drug out; some carry enzymes that break the antibiotic apart; some alter the target so the antibiotic no longer binds. When treatment begins, the susceptible bacteria die in their millions, and any bacteria with protective traits become a larger fraction of the remaining population by simple arithmetic. If those survivors then reproduce, the infection rebounds in a new form: not merely the same bacterium again, but the bacterium *after* the sieve.

The brutality of this is easiest to grasp in places designed to be sterile. A hospital does not merely contain sick people; it concentrates antibiotics, disinfectants, invasive procedures, and vulnerable hosts. For bacteria, it is an obstacle course with prizes at the end. Palumbi's influential essay on humans as an evolutionary force (2001) framed this as a central feature of the modern world: we are not only changing the *ecology* in which organisms live, but actively redirecting their *evolution*. In that sense, a hospital is not just a building. It is an evolutionary environment as distinctive as a coral reef or a desert.

The outcome is not always the dramatic image of a "superbug," but the mundane reality of treatment failure. A urinary tract infection that once responded to a standard antibiotic now requires a second-line drug; a post-surgical infection lingers; a newborn's sepsis becomes harder to manage. The evolutionary point is simple and unsettling: every time an antibiotic successfully kills susceptible bacteria, it also increases the relative representation of any resistant variants that happen to be present. We can be grateful for the lives saved while still acknowledging the selection pressure created. The two facts are not in conflict; they are two sides of the same mechanism.

Cities, meanwhile, create a different suite of kill zones. Urban environments are often hotter than surrounding rural areas because concrete and asphalt store heat and release it slowly at night, a phenomenon known as the *urban heat island.* Cities also fragment habitat into patches, change food availability, alter predator communities, and bathe organisms in artificial light and chronic noise. Some of these pressures kill; others mainly disrupt reproduction or behaviour. Either way, they bias who breeds.

One of the quiet revelations of recent research is that cities can generate evolutionary changes not only in pests but in ordinary native species living alongside us. White clover, for example, carries a defence trait called cyanogenesis: when damaged, some plants can release cyanide as a deterrent to herbivores. The trait is genetically controlled and has costs as well as benefits. Large-scale studies led by Marc Johnson and collaborators, and later global analyses by James Santangelo and colleagues, have shown that the frequency of cyanogenic clover can shift along urban-to-rural gradients, often decreasing in city centres. The details vary among cities, but that variability is part of the story: "urban" is not one environment. A snowy city where winter ground temperatures fluctuate and insulation from snow cover is reduced can impose different selection than a milder city where drought stress dominates. In other words, humans do not build a single new world. We build many, and evolution responds locally.

Urban kill zones also include ones that are not obvious until you think in terms of daily life. Road networks, for instance, do not merely move cars; they kill animals. For species that attempt to cross, the mortality can be high enough to favour behavioural shifts: reluctance to cross open ground, altered activity times, changes in dispersal. For some organisms, this is purely a matter of learning within a lifetime. For others, if the behavioural tendency is heritable, the road becomes a selective agent. The city then shapes not just where a species can live, but what kind of individuals it produces.

If cities and farms show the creative side of human selection—new adaptations, new trait combinations—fisheries show the same logic in a starkly economic light. Many fisheries remove the largest individuals preferentially, either because nets are selective for size or because regulations protect juveniles and target adults. That seems sensible: you want fish to have a chance to reproduce before capture. Yet the consistent removal of large, late-maturing adults can create selection for fish that mature earlier at smaller sizes. In the language of life-history theory, the mortality schedule has been changed: death is now more likely later in life, not because of predators, but because of hooks and trawls. Over time, the population can shift towards "faster" lives. This can reduce yield and resilience, and it can persist even if fishing pressure is reduced, because evolution does not instantly reverse when the filter is relaxed.

The historical record gives a vivid case study of a human-made filter operating in plain sight: the rise of industrial melanism in the peppered moth in Britain. Before heavy industrial pollution, pale, speckled moths were well camouflaged on lichen-covered trees. As soot darkened bark and lichens died back, darker moths gained a survival advantage because birds found the pale ones more easily.

This is often told as an almost mythic parable of Darwinian selection you can see with your own eyes, but the modern molecular work deepens rather than diminishes the story. Research by Arjen van't Hof, Pascal Campagne, and colleagues (2016) showed that the melanic form was caused by the insertion of a transposable element near a gene involved in wing patterning. That is, the dramatic shift in moth colour was not an airy abstraction; it traced to a specific genetic event that rose quickly under a specific human-made condition. The air of industrial Britain became not just a pollutant, but a selective environment.

What makes the "human asteroid" analogy so powerful is that it captures both catastrophe and creativity. Many species, perhaps most, cannot adapt fast enough to keep up with the pace of habitat loss, climate shifts, pollution, and exploitation. For them, the filter is too narrow; the population cloud does not fit through at all. The global assessment work of IPBES has emphasized the scale of biodiversity decline driven by land and sea use change, direct exploitation, climate change, pollution, and invasive species. Those drivers do not operate as gentle nudges. They operate as major, sustained changes to survival and reproduction. Extinction, on this view, is not evolution's opposite; it is what happens when selection is so strong, or the environment so altered, that adaptation cannot rescue the population.

But for those that do persist, human pressure can open doors. Consider organisms that thrive on our waste streams or our built structures: insects in sewers, microbes in plumbing systems, birds nesting on buildings, plants colonizing pavement cracks. These are not necessarily "winners" in any moral sense, but they reveal something important about evolution's opportunism. A new habitat created by humans can be as real an ecological opportunity as a volcanic island emerging from the sea. The difference is that our habitats appear everywhere, and they come with repeating hazards: heat, toxins, and disturbance. The organisms that prosper are those that can both exploit the opportunity and withstand the hazards.

A final twist is that humans rarely apply a single filter at a time. Natural selection in the real world is a messy tug-of-war among pressures that can reinforce or cancel one another. A city may be hotter, favouring heat tolerance, but also fragmented, reducing movement and gene flow, which can limit the spread of beneficial variants. A farm may spray pesticide, favouring resistance, but also rotate crops, which changes the ecological context and can slow the spread of any one resistant type. A hospital may impose antibiotics, favouring resistance, but also enforce infection control measures that reduce transmission. The evolutionary outcomes depend not only on the strength of a single pressure but on the pattern created by many pressures interacting.

The human asteroid, then, is not one blow but a sustained reshaping of the selective landscape. It makes evolution visible on the timescale of a news cycle: a weed that no longer dies when sprayed, a bacterium that ignores a formerly reliable drug, a fish stock that becomes smaller and earlier-maturing, an urban plant population whose defences shift along city streets. These are not merely stories about organisms "coping." They are demonstrations that heredity, variation, and differential survival are always present, waiting for the environment to make their consequences obvious. In the age of human dominance, the environment has become unusually good at doing exactly that.

1.2 Darwin on Fast Forward

Natural selection is often described with the patience of geology: a slow pressure that, given enough millennia, can grind a population into a new shape. That picture is comforting. It suggests that evolution is a distant process, safely separated from human timescales by a moat of deep time. The trouble is that it is only sometimes true.

Natural selection is not a clock. It is a sorting rule. Whenever individuals differ, and those differences affect who survives long enough to reproduce, the population can change in the very next generation. The only reason evolution *seems* slow in many familiar cases is that the sorting is weak: small differences in survival, small differences in family size, and a lot of noise from weather, luck, and chance encounters. Turn up the stakes—make the difference between trait A and trait B a matter of life or death—and the pace can become startling.

A useful way to keep one's footing is to translate the word *selection* into something more concrete: *who gets to be an ancestor.* That is the real currency. Many organisms live; fewer reproduce; fewer still contribute descendants far into the future. Selection is strong whenever a particular trait sharply increases the odds of becoming an ancestor. It is weak whenever almost everyone reproduces regardless of their traits.

One can see the logic with a simple thought experiment that avoids genetics at first. Imagine a population in which 10% of individuals have a trait that helps them survive a new hazard—say, an insecticide in a field, or a novel pathogen. If the hazard kills half the population indiscriminately, the trait will not become much more common; it helped no one in particular. But if the hazard kills 90% of those without the trait while sparing most of those with it, the survivors are no longer a random sample. They are biased toward the trait. Even before anyone has children, the population has effectively been "edited." When the survivors reproduce, the next generation begins life enriched for the trait, and what looked like a rare quirk can become a defining feature in a handful of generations.

This is the engine behind the feeling that evolution is happening "in real time" in human-dominated environments. The environments are not merely new; they are harsh in a specific, repeatable way. The difference between being resistant and being susceptible can be the difference between leaving descendants and leaving none. Under those conditions, natural selection behaves less like a sculptor slowly refining a statue and more like an emergency system that reroutes a population toward whatever configuration can keep it functional.

It helps to put a small amount of mathematics on the table, not to intimidate, but to give intuition a spine. Biologists often summarize selection strength with a number called a *selection coefficient*, usually written as s. In plain language, s captures how much worse one type does compared to another. If a genotype has $s = 0.1$, it is doing about 10% worse in the relevant currency (survival to reproduction, or number of offspring). If $s = 0.5$, it is doing 50% worse. Those are radically different worlds.

Now connect this to time. Suppose a beneficial trait begins at frequency p in a population and has a strong advantage. The details depend on dominance, mating patterns, and chance, but the general rule is simple: when selection is strong, the frequency curve is steep. It can go from "hardly seen" to "common" within tens of generations, and for organisms with short generation times, tens of generations can be a few years.

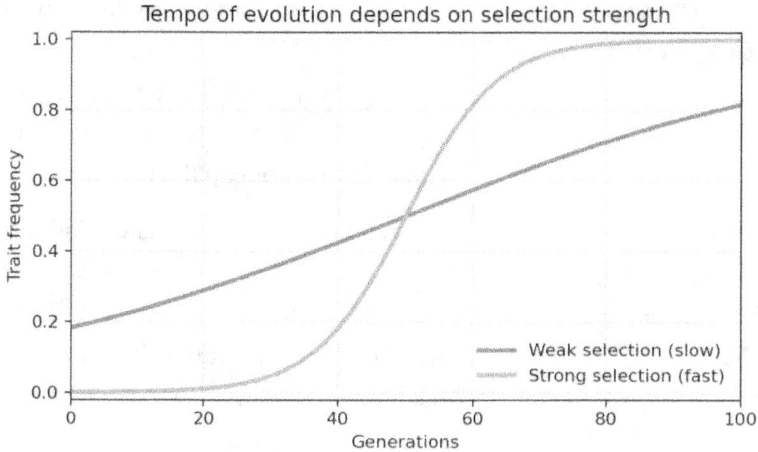

The figure is deliberately simple: it does not assume any particular gene, only that the trait is heritable and that selection is either weak or strong. The key point is not the exact curve shape but the difference in slope. The slow line spends a long time hovering in the middle, changing gradually. The fast line spends very little time there at all; it rises quickly because the "losing" type is being removed efficiently. A population in such a regime is not evolving by tiny nudges. It is being pushed.

The most persuasive demonstrations of fast evolution do not come from the laboratory, where conditions can be controlled but critics can complain that nature is messier. They come from field studies in which nature is indeed messy and yet evolution can still be detected, predicted, and even induced.

Trinidadian guppies provide a classic example. In the rivers and streams of Trinidad, guppies live under different predation regimes. Some pools contain predators that strongly target adult fish, while other pools are comparatively safe for adults. Decades of work, including influential studies by John Endler and David Reznick, connected these predation patterns to consistent differences in guppy life history: when adult mortality is high, there is an advantage to reproducing earlier and investing heavily in reproduction, because waiting is risky. When adult mortality is low, it can pay to grow larger and delay reproduction, because there is time to reap the benefits of size.

So far, this might sound like an ecological story: predators influence behaviour and life history, full stop. The deeper question is whether those differences are merely flexible responses within a lifetime or whether populations have actually evolved. Reznick and Bryga's transplant experiment (1990) addressed this with the bluntness of an engineer. They moved guppies from a high-predation environment into a low-predation environment and then tracked changes over time. Within about 11 years—on the order of 30 to 60 guppy generations, depending on conditions—the introduced populations shifted in the predicted direction: later maturity, changes in reproductive effort, and other life-history traits associated with lower adult mortality. The point is not that guppies are magical. It is that the selection pressure was strong and consistent enough to leave an evolutionary signature on a timescale that fits comfortably within a human career.

Why does predation produce such speed? Because it is not a gentle preference. It is a filter with teeth. If a certain life history increases survival to reproduction under a particular mortality schedule, those individuals become disproportionately represented among the ancestors of the next generation. Once the environment changes, a different life history becomes the better bet, and the ancestry of the population shifts accordingly. Natural selection, in this setting, is not waiting for eons. It is reacting, generation by generation, to who gets eaten.

Industrial melanism in the peppered moth offers a different kind of clarity, because it unfolded against a historical backdrop that is almost too neat: the soot-darkened landscapes of the Industrial Revolution and the later cleanup of air pollution. The broad outline is familiar. In the early nineteenth century, as coal smoke blackened trees and killed lichens, darker moths became better camouflaged, and their frequency rose. Later, as air quality improved, the advantage reversed and lighter moths became more common again. This is natural selection with a toggle switch.

What makes the peppered moth story especially valuable for a modern reader is that it links visible change to molecular mechanism. Van't Hof, Campagne, and colleagues (2016) traced the classic dark "carbonaria" form to a transposable element insertion in a gene involved in wing patterning. They also inferred that the mutation likely arose around 1819, aligning the genetic event with the historical period when the selective environment changed most dramatically. The soot did not "cause" the mutation; mutations arise without regard to need. But the soot made one rare mutation matter enormously. When the environment becomes a kill zone for the common type, a previously insignificant variant can become the winning ticket.

This interplay between chance and necessity can be surprising. People often imagine evolution as purposeful, as if populations "try" to adapt. In reality, populations carry a library of small differences, most of which are neutral or nearly so under ordinary conditions. Strong selection is what makes that library suddenly relevant. It takes a page that was gathering dust and turns it into the main text.

The finches of the Galápagos have a similar power to make selection feel immediate, partly because the stakes can be counted in beaks and seeds rather than in genes and enzymes. Peter and Rosemary Grant's long-term work on Darwin's finches documented episodes in which drought altered the seed landscape, changing which beak sizes could efficiently feed. In one later genomic analysis, the Grants and collaborators (2016) connected beak size differences to a region containing the gene $HMGA2$, and reported selection that could be extraordinarily strong in particular years, with an estimated selection coefficient on the order of $s \approx 0.59$ against certain genotypes. Numbers like that are not gentle. They imply that, in that context, having the "wrong" version is close to a reproductive disaster.

One year of severe selection does not guarantee permanent change. A wet year can reverse the pressure; immigration can bring in other variants; randomness can blur the outcome in small populations. Yet the broader lesson stands: when ecological conditions swing and the link between trait and survival becomes tight, evolution can proceed at a pace that looks more like crisis management than gradual artistry.

The mathematics also clarifies why some traits respond quickly while others seem stubborn. If a trait is governed by many genes, each with small effects, the response can be slower, because selection must sift through a subtler pattern. If a trait has a simple genetic basis, selection can change it rapidly, because there are fewer genetic "degrees of freedom." If the beneficial variant is already present at a reasonable frequency, change can be swift; if it begins as a single rare mutation, chance has more opportunities to erase it before it spreads. And if the environment changes faster than a generation time, even strong selection may not help, because the population cannot "update" quickly enough. The emergency system works only at the speed of reproduction.

Human activity often supplies two ingredients that make fast evolution more likely. One is intensity: pesticides, antibiotics, and harvesting can produce very high mortality that depends on specific traits. The other is predictability: the same filter is applied repeatedly, season after season, prescription after prescription. For a population, a predictable hazard is a stable target. Once a resistance trait appears, it is rewarded again and again.

At the same time, there is a reason not to romanticize speed. Fast evolution is not synonymous with improvement in any broad sense. It is simply a rapid increase in whatever traits help organisms leave descendants under current conditions. In a pathogen, that might mean drug resistance. In a weed, herbicide resistance. In a fish stock, earlier maturation at smaller size, which can reduce both yield and ecological function. Even when the evolutionary change is impressive, it can be costly. Resistance mechanisms often carry tradeoffs: pumping toxins out of cells consumes energy; thickening a cell wall can slow growth; altering a target molecule can impair normal function. Under strong selection, those costs are tolerated because the alternative is death. When selection relaxes, costs can matter again, sometimes allowing susceptible types to return. Evolution is opportunistic, not sentimental.

There is also a subtle psychological trap in watching evolution speed up: it can create a false sense of control. If populations can evolve quickly, perhaps they can always evolve quickly; perhaps nature will simply adapt to whatever humans do. That comforting thought is wrong in the way that a half-truth is wrong. Some populations do evolve fast; others do not. Even for those that can, there are limits set by genetic variation, population size, and the sheer brutality of change. A filter that kills 99% of individuals does not merely select; it risks pushing the population into collapse. In small remnants, chance can overpower selection, wiping out beneficial variants before they can spread. Evolution can rescue a population, but it can also fail, and failure can look like silence: no more calls at dusk, no more insects on the windshield, no more fish on the line.

The most honest way to picture natural selection on fast forward is as an emergency response that is sometimes effective and sometimes overwhelmed. It does not plan ahead. It does not preserve what we value. It simply amplifies what survives. Under ordinary conditions, the emergency system is rarely triggered; the changes from one generation to the next are modest, and the long arc of evolution can look smooth. Under crisis conditions—a sudden predator regime, a soot-darkened forest, a drought that reshapes the available food—the system activates. The population's genetic makeup is not revised by intention but by attrition, and the future is written by whoever remains to reproduce.

On a windless day in an industrial town, with soot settling invisibly on bark and stone, the selective environment is already being rewritten. In a stream pool where predators patrol, the timetable of reproduction is being revised one generation at a time. In a drought-struck island where only certain seeds can be cracked, the difference between one beak shape and another becomes a difference between ancestors and non-ancestors. The mechanism is the same in each case, as spare and unsentimental as arithmetic: survival is not evenly distributed, and heredity makes the imbalance accumulate.

1.3 The Illusion of Change

A strange thing happens whenever people begin to notice evolution up close: the word expands to fill everything that changes. Birds sing differently in noisy streets, so they must have evolved. City trees leaf out earlier, so they must have evolved. Fish are getting smaller, so they must have evolved. Sometimes those statements are correct. Often they are not. They may be describing something just as real and sometimes just as impressive, but it is a different category of biological response: *phenotypic plasticity*.

The term sounds technical, yet the idea is familiar. A phenotype is an organism's observable form and function: its body size, colour, timing, metabolism, behaviour, immune profile. Plasticity means that the same individual, with the same DNA, can produce different phenotypes under different conditions. The human body is a plastic system. At high altitude, many people make more red blood cells. Under strength training, muscles grow. After infection, immune memory changes. None of that requires a genetic mutation. It is the body adjusting its settings.

Plasticity is neither fake nor trivial. It is a major reason organisms survive in variable environments. It can be the difference between life and death in a bad season. But plasticity is not the same as genetic evolution, because it is not, by itself, a change in inherited instructions across generations. Genetic evolution happens when DNA variants become more or less common in a population because they influence who leaves descendants. Plasticity happens when a single genotype can produce different outcomes depending on the environment.

The two processes can interact. Plasticity can buy time for a population to persist long enough for genetic adaptation to occur. It can also mask genetic change, making evolution harder to detect. And sometimes plasticity itself evolves: some genotypes may be better at flexing than others, and selection can favour that flexibility. The relationship is subtle. The illusion of change arises when we look at a shifting phenotype and assume, without evidence, that genes have shifted too.

This is not merely a semantic issue. It is a practical one. If a trait shift is mostly plastic, it may reverse quickly if conditions improve. If it is genetic, it can persist even after the original pressure is removed. If we mistake plasticity for evolution, we may overestimate nature's capacity to "keep up" with human change. If we mistake evolution for plasticity, we may miss long-term consequences of our actions, including changes that are hard to undo.

A clean way to draw the distinction is with a question that sounds almost moral: *what gets passed on?* Plasticity changes the individual.

Evolution changes the population's inherited makeup. The difference sits inside the word *heritability*. In everyday speech, heritability sounds like a simple yes-or-no property: is a trait inherited or not? In biology it has a narrower meaning. Heritability asks what fraction of the variation in a trait, in a particular population in a particular environment, is associated with genetic differences among individuals. That is why it can change with context. A trait can be "highly heritable" in one setting and less so in another, not because genes are fickle, but because environments can swamp or reveal genetic differences.

The danger is that the world is full of confounders: alternative explanations that can mimic evolution. A population may become smaller not because individuals evolved to grow less, but because food became scarcer. A city population may differ from a rural population not because of adaptation but because only certain individuals disperse into cities and survive there, a kind of demographic sorting. Or the city population might simply be founded by a small, unrepresentative sample of the rural population, allowing chance (*genetic drift*) to change gene frequencies without any adaptive advantage. In fragmented landscapes, gene flow may be reduced, letting populations diverge for reasons that have nothing to do with the specific urban stressors we suspect.

Marina Alberti and Andrew Hendry (2017) argued persuasively that cities can influence nearly every evolutionary force at once: selection, drift, gene flow, mutation, and demographic change. That breadth is part of what makes urban evolution fascinating. It is also what makes it easy to tell a confident story for the wrong reason. When many processes operate together, a visible difference between populations does not automatically diagnose adaptation.

The most common illusion comes from taking a "before and after" observation and treating it as a genetic verdict. The fish are smaller now than they were thirty years ago, therefore evolution. The birds are bolder now, therefore evolution. But "after" populations differ from "before" populations in multiple ways. The environment has changed, which can alter growth and development. The age structure can change: if older individuals are removed, the average size drops even if the growth rules have not changed. And the population might now contain a different mix of individuals because the vulnerable ones died or moved away. That last point is particularly slippery: selection within a generation can change the *composition* of the survivors without having yet changed the *genetic composition* of the next generation. It is sorting, yes, but it becomes evolution only if the survivors' traits are heritable and they pass them on disproportionately.

The cleanest way to avoid the illusion is conceptually simple and logistically hard: raise organisms from different environments in the same conditions and see whether their differences persist. This is the logic of a *common garden* experiment. If city and rural plants look different in the wild but become similar when grown side by side in identical soil and light, the difference was largely plastic. If they remain different, genetics is implicated. A related approach is a *reciprocal transplant*, moving organisms into one another's environments to test whether locals outperform foreigners, a sign of local adaptation.

Even these designs have pitfalls, and Juha Merilä and Andrew Hendry (2014) laid out several reasons why rapid trait change in nature is often hard to interpret. Phenotypes can shift for many reasons. Even if there is genetic change, it may not be adaptive; chance can move genes around too. And even if the change is adaptive, it may not be caused by the driver we think it is. Climate change is an obvious suspect for many shifts, but land use, pollution, species interactions, and management practices can change at the same time. Untangling them demands careful design, replication, and humility.

A second common illusion comes from the popularity of tidy narratives. People like origin stories with a clear start date: "this population evolved after the subway opened" or "this trait appeared once humans arrived." The real histories are often older and messier. A celebrated example concerns the mosquito form known as *Culex pipiens* f. *molestus*, often associated with underground urban habitats and famous in public storytelling as a "subway mosquito" that supposedly evolved rapidly in modern transit systems. The details vary depending on the retelling, but the emotional core is irresistible: the city builds tunnels, and within a century a new mosquito emerges, biting mammals rather than birds, breeding year-round in warm subterranean air.

The trouble with irresistible stories is that they can outrun evidence. Population genomic studies have suggested that the "underground" mosquito lineage in some regions is far older than modern subways and likely tied to longer histories of human-associated habitats. The precise timing and geographic origins remain topics of investigation, but the broader lesson is robust: a population can look perfectly tailored to a modern niche while having deeper roots in older human environments such as basements, sewers, wells, and storage spaces. In that case, the rapid-change narrative is an illusion created by the modern observer's calendar rather than by the organism's evolutionary history.

The illusion of change also appears in reverse: sometimes evolution is dismissed because the phenotype can change quickly. If a trait shifts within a lifetime, people may assume it cannot be genetic. But populations can undergo genetic change even while individuals remain flexible. Think of a thermostat. A house can still heat up and cool down in response to weather, even if you reprogram the thermostat's default settings. Similarly, a species can remain plastic while the genetic baseline shifts. The result is that simple observation of plasticity does not exclude evolution; it merely means one must test.

The conceptual difference between plasticity and adaptation is easier to see with a deliberately stripped-down cartoon.

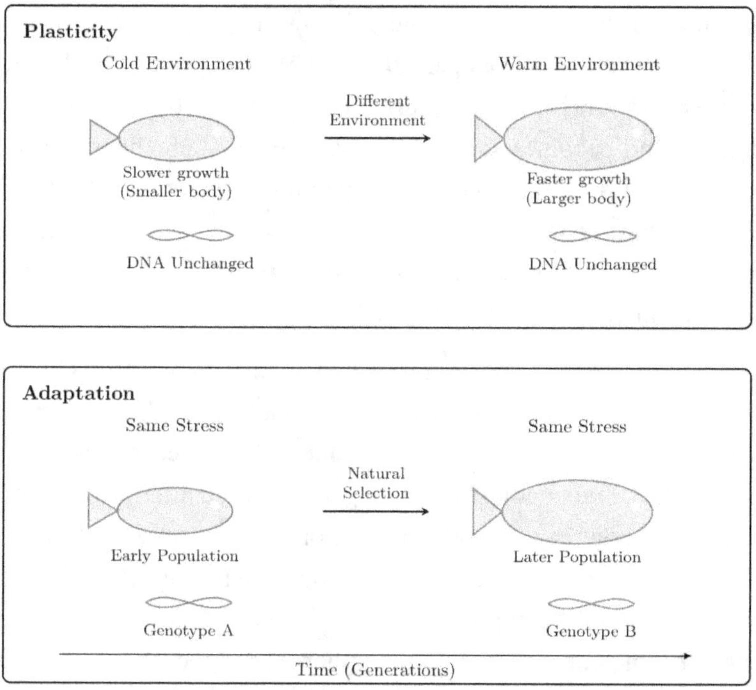

Plasticity: The same DNA produces different phenotypes in response to different environments. **Adaptation:** Over generations, the genetic baseline of the population shifts (e.g., Genotype A is replaced by Genotype B), leading to permanent phenotypic changes even if the environment remains constant.

The top panel shows a single genotype producing different outcomes because development responds to temperature or food. The bottom panel shows something slower and less visible: the DNA variants in the population shift over generations, so that even if the environment were held constant, the population's typical phenotype would change.

A concrete example helps because the distinction can feel abstract until it is pinned to a familiar kind of claim. Consider the recurring observation that many harvested fish populations have become smaller and mature earlier. This could be genetic adaptation to size-selective fishing, and in some cases there is evidence pointing in that direction. But the same pattern can also be produced without genetic evolution. If fishing removes older, larger individuals, the remaining population becomes younger on average. A younger population is smaller on average, even if individuals grow exactly as they always did. If food becomes scarce because of ecosystem change, growth can slow. If densities change, competition can change. A "smaller fish" headline can therefore describe demographic structure, environmental effects on growth, or genetic evolution—and without careful work it can be difficult to say which.

The same ambiguity shows up in urban environments. Birds in cities often sing at higher frequencies, presumably to be heard over low-frequency traffic noise. Is that evolution? Sometimes it may be. But birds can also adjust their song within their lifetimes, learning from neighbours and responding to noise conditions. A shift in average song frequency across a city could therefore reflect learning and behavioural flexibility rather than genetic change. Even if there is a heritable component to singing behaviour, the observed change might be dominated by plasticity. The phenotype is real either way, but the causal story differs.

Why does it matter which story is true? Because policy and prediction depend on it. If a population is coping mainly through plasticity, it may be living close to its limits. Plastic responses are often costly; they can involve stress hormones, diverted energy, reduced reproduction, and trade-offs that do not show until later. Plasticity can keep the organism upright while slowly draining its reserves. A population that looks "fine" because individuals can flex might still be declining in fertility or survival in ways that are easy to miss until numbers crash.

Conversely, if the response is genetic, it may not be easily reversible. A pesticide-resistant insect population can remain resistant long after the pesticide is removed if the resistance carries little cost, or if compensatory evolution reduces the cost. A fish population that has evolved earlier maturation may not spring back immediately when fishing stops, because the evolved life history has become the new normal. In those cases, the illusion is not that change happened, but that it is a simple, short-term adjustment.

The scientific community has become increasingly candid about how often popular narratives leap over these complexities. Santangelo and Johnson (2021), reviewing the burgeoning literature on adaptive evolution in cities, argued that genuinely comprehensive demonstrations of *adaptive* urban evolution are rarer than the sheer number of published studies might suggest. The phrase "urban adaptation" is sometimes used as shorthand for "urban difference," even when alternative explanations are not ruled out. This is not cynicism; it is scientific hygiene. When a field becomes fashionable, weak evidence can be amplified simply because people are looking for it.

There is also a human temptation to treat any trait change as proof that nature is resilient, that life will find a way around every obstacle. Plasticity encourages that temptation because it can produce rapid shifts that look like instant adaptation. A plant grows smaller leaves in drought; an animal changes its activity schedule; a microbe alters its gene expression when exposed to stress. These are genuine feats of biology. Yet they may be the equivalent of holding one's breath rather than evolving gills.

The most convincing studies therefore tend to combine multiple lines of evidence: a demonstrated trait difference, a demonstration that the difference persists in a common environment (implicating genetics), evidence that the trait affects survival or reproduction (showing it matters), and increasingly, genomic signals consistent with selection. This is why large, replicated datasets can be so powerful. The global white clover work led by Santangelo and colleagues (2022), for example, combined enormous sampling with genetic and genomic analyses, strengthening the case that at least some urban—rural trait clines reflect adaptation rather than mere plasticity or chance. Even then, the finding is not that cities always produce the same evolutionary answer, but that they create many local experiments, some of which run strongly enough to leave a genetic mark.

A final source of illusion is more philosophical: the tendency to mistake a good fit for a new origin. When an organism seems perfectly suited to a human environment—a mosquito thriving underground, a weed shrugging off herbicide, a bacterium persisting on a hospital surface—it is natural to imagine that the trait must have arisen in response to the modern setting. Sometimes it did. Sometimes the trait is older, co-opted from a different context, or assembled from preexisting variation. Evolution does not always invent from scratch; it often repurposes. The apparent novelty can be an artifact of our attention: we notice the organism only once it becomes a nuisance.

The cure for the illusion of change is not to become suspicious of every claim of rapid evolution. It is to learn a more discriminating excitement. Plasticity is thrilling in its own right: it reveals the organism as a responsive system, capable of multiple stable states. Evolution is thrilling because it reveals how populations can be reshaped by differential survival. Both processes matter in a human-altered world, and both can occur together.

When a city's summers become hotter, some individuals will cope by physiological adjustment, some by behavioural change, some not at all. When a pollutant appears, some organisms will switch on detoxification pathways, others will fail, and over time the balance of inherited variants may shift. The visible outcome may be the same—a population that looks different from before—but the machinery beneath it determines whether the change is a fleeting posture or a lasting rewrite.

A good rule of thumb is to treat any single observation of a shifted phenotype as the beginning of a question, not the end of one. The question is not "did it evolve?" but "what combination of flexibility, sorting, chance, and inheritance produced what we see?" That combination decides whether the change will vanish with the first good year, or persist as an echo of the bad ones long after the original stress has faded.

Chapter 2

The Shape-Shifter's Toolkit

Evolution requires raw materials; natural selection cannot sculpt from a void. By examining the hidden libraries of genetic variation and the chaotic influence of chance, we uncover the mechanics that allow species to transform with startling speed when the clock is ticking.

2.1 The Warehouse of Variance

Imagine a population as a workshop that is suddenly handed a new job. Yesterday it made wooden chairs; today the order is for steel scaffolding, due tomorrow morning. One way to cope is to wait for a worker to invent welding from scratch. The other is to remember that someone in the back already knows how to weld, someone else has been hoarding scrap metal, and a third has a book of diagrams. Evolution, when it moves fast enough to notice within a human lifetime, usually resembles the second scenario. It does not begin by asking for novelty. It begins by *sorting what already exists.*

The phrase *standing genetic variation* sounds technical, but it points to something ordinary: individuals differ. They differ in ways you can see (size, color, tolerance to cold), and in ways you cannot (how tightly a drug binds to a protein, how quickly a detox enzyme breaks down a pesticide). Some of these differences matter every day; many make almost no difference until the environment changes. A population carries, in its collective genome, a warehouse of variants: some common, some rare, some silent for decades, some harmful in one setting but helpful in another.

This warehouse matters because mutation is slow in the way that matters most for survival: it is slow *on demand*. New mutations arise continually, but they arrive as a trickle, not as a targeted delivery. If the environment shifts abruptly—a new antibiotic, a drought, an introduced predator, a toxic pollutant—the population does not get to request the precise mutation it needs. It must make do with whatever variants are already on the shelves.

That constraint produces an evolutionary speed limit. A back-of-the-envelope version goes like this. Suppose a single genetic change would make an organism much more likely to survive a new toxin. New mutations that specific are rare. Even in huge microbial populations, the right change might not appear immediately in the right gene, on the right background, in the right place. In small populations—an endangered fish in a warming river, an isolated alpine plant—waiting for that one helpful change can be like waiting for a particular snowflake in a blizzard. Most will melt on arrival. Meanwhile, selection is not a gentle judge. It is a sieve, and when the sieve suddenly tightens, a population can be reduced to a few survivors before the helpful mutation even appears.

Standing variation breaks this bottleneck by shifting evolution from *invention* to *retrieval*. When an environmental change makes a previously rare variant valuable, selection can increase its frequency immediately, in the very first generation after the change. The allele was already there, perhaps at 1%, 0.1%, or even lower. The critical point is that it existed in multiple individuals, not as a single new copy. Instead of waiting for the winning lottery ticket to be printed, the population discovers it already sold a handful, tucked into wallets across town.

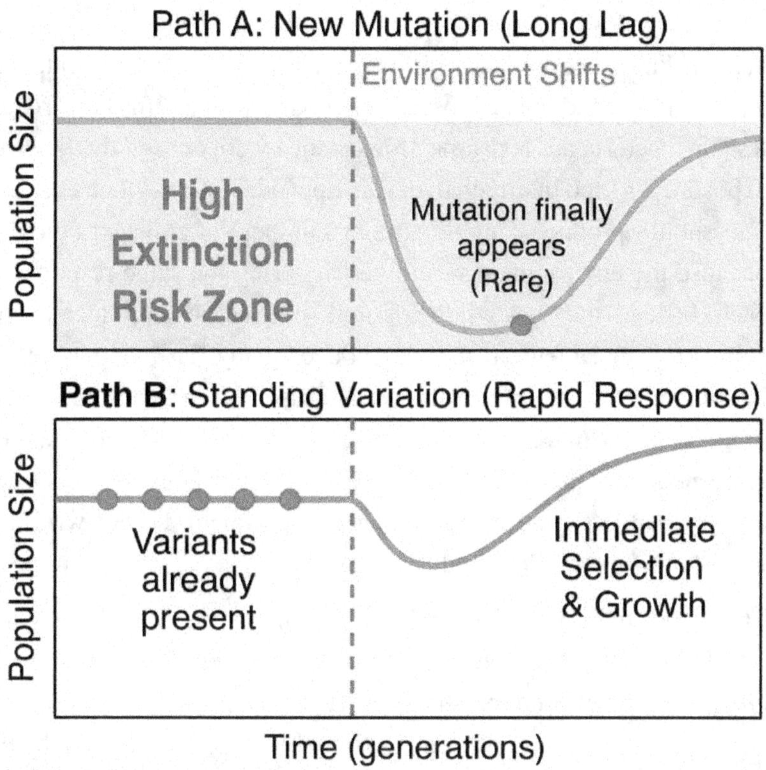

The contrast in the diagram is not merely aesthetic. It captures a difference between two biological clocks. Mutation keeps time in the currency of *chance*: when will a particular genetic change occur? Standing variation keeps time in the currency of *sorting*: once a change in circumstances makes a variant valuable, how quickly can it spread? Sorting can be fast because selection is fast. If a resistant individual leaves twice as many offspring as a susceptible one, the arithmetic compounds with each generation. What slows evolution, so often, is not selection's strength but selection's lack of raw material.

A classic historical anecdote makes the point with unusual clarity. In the early 1940s, Salvador Luria and Max Delbrück devised what became known as the fluctuation test. They grew many separate bacterial cultures and then exposed them to a virus that kills bacteria unless the bacteria are resistant. If resistance were *caused* by the virus exposure—a kind of directed, on-the-spot adaptation—then each culture should produce roughly similar numbers of resistant colonies, because the environment would be triggering the same response in each. Instead, they saw wild variation: some cultures produced none, some produced a few, and some produced many. The simplest explanation was that resistance mutations arose *before* exposure, at random times during growth. A culture in which a resistance mutation happened early produced many resistant descendants; one in which it happened late produced few. The environment did not write the mutation; it merely revealed it.

That experiment is often taught as a triumph of genetic reasoning, which it is. But its deeper message is about tempo. When conditions change, survival often depends on what was already present in low numbers. Even if the key variant originally arose by mutation, by the time the crisis arrives it is no longer a single new event. It is part of the standing variation, scattered through the population like seeds.

The same logic applies far beyond bacteria and viruses. Consider antibiotics in hospitals, pesticides in agriculture, or herbicides sprayed across a field. These are among the most intense, sudden selection pressures humans have ever created, and they have the bluntness of a guillotine: most susceptible individuals die, and the few survivors repopulate. When resistance evolves rapidly—sometimes startlingly rapidly—it is often because resistant variants were already present at low frequency. They might have been mildly costly before the drug was introduced, keeping them rare; they might have been neutral, drifting quietly; or they might have been maintained by a patchwork of environments in which the drug was sometimes present and sometimes absent. Once the drug becomes common, the fitness accounting flips. A cost that mattered yesterday becomes irrelevant compared to the benefit of staying alive.

A more intimate version of this story plays out in our own bodies. Human immune genes are famously variable, and that diversity is not an evolutionary accident. Pathogens change quickly; immune systems face a moving target; and a population with a broader repertoire is harder to exploit. Some of that variation is maintained by a form of selection that does not push one "best" allele to fixation. Instead, it keeps multiple alleles in circulation because rarity can be an advantage (pathogens adapt to the common types), or because different alleles are favored in different ecological or disease contexts. This is one way a species can stock the warehouse for future battles.

Standing variation is not always obvious, because much of it is *cryptic*: differences that have little effect until the environment changes. A protein might be slightly more stable at high temperature; a membrane might leak a bit less in salty water; a detox enzyme might be a touch more efficient against a compound that the organism rarely encounters. Under ordinary conditions, these differences can be nearly invisible. Under stress, they suddenly matter. Cryptic variation is why populations can sometimes respond to a new challenge with surprising speed, as though the solution was hiding in plain sight.

To see how quickly selection can work when the warehouse is well stocked, it helps to focus not on a single dramatic gene but on many modest contributors. In real organisms, traits like heat tolerance, body size, or resistance to a toxin often involve many genes. Each may contribute a small effect, and the combined outcome determines survival. When selection acts on such a trait, evolution can proceed through shifts in frequencies at many loci at once. The population does not need to find one miraculous mutation. It can climb uphill by rebalancing what it already has.

This is one reason why the genomic "signature" of rapid adaptation from standing variation can look different from the textbook image of a single new mutation sweeping through a population like a brush fire. In the older, simpler picture, a beneficial mutation appears once and rises quickly to fixation, dragging along neighboring DNA sequences because of genetic linkage. That creates a strong, distinctive pattern: a reduction in genetic diversity around the selected site. But when adaptation draws on standing variation, there may be multiple copies of the beneficial allele already present on different genetic backgrounds, or multiple alleles with similar effects. Selection then increases several lineages in parallel. Genetic diversity is reduced less severely, and the sweep is "soft" rather than "hard." Barrett and Schluter (2008) emphasized how often rapid adaptation should be expected to follow this softer, faster route, simply because nature rarely begins from a single copy.

Seasonal environments offer a particularly vivid demonstration of the warehouse principle, because the direction of selection can change back and forth on a timescale you can mark on a calendar. Fruit flies provide a celebrated example. Bergland and colleagues (2014) reported that many genetic variants in *Drosophila* shift in frequency across seasons—rising in one part of the year, falling in another. The striking implication is not only that selection can be strong, but that genetic variation can be maintained over long stretches of time precisely because conditions fluctuate. The warehouse is not a dusty storeroom; it is a rotating inventory system. Alleles do not need to be permanently "best" to persist. They need only to be useful often enough, or in enough places, to avoid being eliminated.

Climate change adds urgency to this discussion because it changes environments quickly and repeatedly, and because it often does so in ways that are hard to escape. A population cannot migrate out of a warming ocean if it is tied to a coastline or a reef. In such cases, the difference between having relevant standing variation and lacking it can be the difference between persistence and collapse. Experimental work on marine organisms has shown how selection can act on pre-existing genetic differences under stressors like acidification, with measurable shifts in survival-related traits and in allele frequencies over short timescales. Bitter and Kapsenberg (2019), for example, examined how larvae respond under different pH conditions, illustrating how the initial diversity present at the outset can shape the scope for rapid adaptive response.

None of this implies that new mutations are irrelevant. Over longer periods, they are the ultimate source of novelty. Even on short timescales, a new mutation can rescue a population that happens to lack useful standing variation. The point is probabilistic rather than absolute: when the environment changes abruptly, populations that can respond tend to do so by amplifying what is already there. Populations that must wait for invention live under a harsher clock.

There is also a social lesson hidden inside this biology, one that becomes uncomfortable when we recognize how often we create the selection pressures ourselves. When antibiotics are overused, when pesticides are applied broadly and repeatedly, when a single herbicide becomes the default across millions of hectares, we are not simply challenging an organism to "evolve resistance." We are sampling its warehouse at high speed. Rare variants become common because we make them valuable. The result can feel like nature is cheating—as though resistance appeared out of nowhere. In reality, it often appeared from somewhere very ordinary: from the low-frequency differences that any large population inevitably carries, waiting for the day they become the difference between life and death.

In this light, the most sobering thought is not that evolution is fast. It is that evolution can be fast *without* being new. The future so often arrives prepackaged, already distributed through the present, and all the environment has to do is change the price tags.

2.2 The Founder's Gamble

A population can change quickly for a reason that has nothing to do with being well adapted. It can change because a door closes behind it.

Picture a storm-tossed raft of vegetation scraping onto an empty volcanic shore, or a handful of insects hitching a ride in the folds of cargo, or a few mice stepping off a ship onto an island whose smells and sounds have never included a predator like them. The founding individuals are not a careful sample. They are whoever happened to be in the wrong place at the right time. When they breed, they do not merely start a new population; they also edit the genetic library they carry with them. Most books never make the trip.

That edit has a name: the *founder effect*. It is a special case of a broader process called *genetic drift*, which simply means random change in the frequencies of genetic variants (alleles) from one generation to the next. Drift happens in all populations, but in large ones it is usually drowned out by the law of averages. In small ones it can roar. If only ten individuals contribute offspring, then chance alone can double the frequency of an allele, wipe another out entirely, or make a rare trait suddenly common. No adaptation is required. No improvement is implied. It is sampling error, made biological.

There is a temptation, when watching a population diverge, to assume that every change is a response to local conditions—that nature is always optimizing, always sculpting. Drift is the counter-story: rapid change without a sculptor. It is why island populations can look strange even before selection has had time to do much. It is why endangered species can lose useful diversity simply because there are too few individuals left to carry it. It is why some outbreaks of pathogens take unexpected genetic turns even when the environment is stable. Chance does not replace natural selection; it competes with it for influence, and in small populations chance can win.

The founder effect begins with arithmetic. If a source population contains many alleles at many genes, the founding group will carry only a subset. The subset will not match the original proportions. A rare allele might not be present at all. Another might be over-represented because two of the founders happen to carry it. This immediate shift can be dramatic, but the deeper consequence is what happens next: a small founding population tends to remain small for a while, and small populations are vulnerable to further drift and to *inbreeding*—mating between relatives—which can expose harmful recessive mutations. The population's early generations become a narrow gate through which the future must pass.

A bottleneck is the same gate closing for an already established population. A fire, a disease, a hunting campaign, a harsh winter, a drought: many events can reduce a population from thousands to dozens. The survivors, like founders, are not necessarily the best adapted; they are often simply the lucky. If the population later recovers in numbers, it may still be genetically impoverished, because the recovery is built from that narrow sample.

The consequences are sometimes visible in historical records. Northern elephant seals were hunted so intensively in the nineteenth century that the species was reduced to a tiny remnant, often described as perhaps a few tens of individuals, before protection allowed numbers to rebound. Today their census size is large, but genetic studies have long pointed to extremely low genetic variation compared with many other mammals, consistent with a severe bottleneck. The seals did not become genetically uniform because uniformity was optimal; they became uniform because most lineages were erased. A recovered headcount can hide a depleted gene pool.

Chance can also manufacture differences between populations that look, at first glance, like local adaptation. If two small groups are founded from the same source and placed in similar environments, they can still diverge rapidly. Each group begins with a different grab-bag of variants. If one happens to lose alleles that influence, say, body size, it may end up smaller on average. If the other keeps them, it may remain larger. A casual observer might attribute the difference to climate or diet, but the initial roll of the dice can be enough.

This is not a minor detail, because early drift can shape what selection is able to do later. Selection cannot act on variants that are missing. If a useful allele is lost during founding, that evolutionary option disappears, at least until it is reintroduced by migration or recreated by mutation. Conversely, drift can push a rare allele to moderate frequency, giving selection something substantial to work with if the environment later changes. Laboratory evolution experiments have repeatedly shown this kind of contingency: replicate populations founded from the same stock, exposed to the same conditions, can adapt along different genetic routes because the early generations drifted in different directions. Even when the end result looks similar, the path can differ.

A concrete case that has fascinated urban biologists involves mosquitoes. Subways, basements, and tunnels create pockets of habitat that are isolated in peculiar ways: warm year-round, humid, rich in standing water, and separated by barriers that matter to a small flying insect. Populations can become trapped in these underground islands, founded by only a few individuals, then multiplying into dense local swarms. A mosquito living and breeding in a tunnel does not need to be "better" than its aboveground relatives to become genetically distinctive. It may simply be descended from a small founding set, shaped by drift, and then prevented from mixing with other populations. Once genetic differences accumulate, selection can add its own touches—differences in biting behavior, host preference, or tolerance to the subway's chemical environment—but the first divergence may be a matter of isolation and chance.

The founder effect has played an outsized role in human stories as well, though with a crucial caveat: humans do not form neat experimental populations, and social history is never reducible to genetics. Still, the population-genetic principle is clear. When a small group splits off and remains relatively isolated, some alleles become common simply because they were carried by the founders. Certain inherited diseases occur at elevated frequencies in particular communities partly for this reason, especially when the founding group was small and marriages tended to be within the community for cultural or geographic reasons. The pattern can be tragic: the same process that preserves identity and continuity can also, unintentionally, concentrate rare harmful variants. Here drift is not an abstraction but a burden measured in clinics and family trees.

It is worth pausing over the phrase *non-adaptive change*. In everyday language, "evolution" is often treated as a synonym for improvement. In biological terms, evolution is merely change in inherited traits over generations. Drift changes traits without regard to whether they help. In small populations it can even push harmful alleles upward in frequency, simply because selection is too weak to purge them reliably when numbers are low. The result can be a decline in average fitness, a phenomenon sometimes called "extinction vortex," and in extreme cases it can contribute to an "extinction vortex," where small size increases drift and inbreeding, which reduce fitness, which makes the population smaller still.

Yet drift is not simply destructive. It is indifferent, and indifference can have creative consequences. By reshuffling allele frequencies, drift can move a population to a new genetic starting point. If the environment changes, that new starting point might prove fortunate. A population that has drifted into an unusual combination of alleles might, by accident, be pre-adapted to a novel condition. This is not adaptation in the moment of drifting; it is luck meeting future circumstance.

One of the strangest and most instructive examples of what small population size can do comes from the biology of islands. Island populations often show a mix of predictable ecological patterns and unpredictable genetic quirks. The predictable side includes shifts in body size and life history that are plausibly shaped by food availability and predation pressure. The unpredictable side is the signature of drift: odd frequencies of color morphs, unusual genetic variants becoming common, rapid divergence between nearby islands that seem environmentally similar. When only a few individuals establish a population, even the most careful ecological explanation should carry a warning label: the starting sample matters.

The mechanisms can be made intuitive with a simple mental model. Imagine a large jar filled with marbles of many colors, representing a genetically diverse population. If you scoop out a handful to start a new jar, the handful's color mix will almost never match the original. If the new jar remains small for several scoops, the mix can drift further, sometimes veering toward a single color. Nothing about the jar rewards one color over another; the outcome is statistical. That is drift.

What complicates the picture, and makes it more interesting, is that genes do not influence traits one at a time. Many traits are shaped by many genes, and genes often interact. When a population is bottlenecked, it does not merely lose alleles; it can also change the amount of heritable variation available for selection in fitness-related traits. Counterintuitively, a bottleneck can sometimes increase the variance in a trait, at least temporarily, by converting hidden genetic differences into expressed ones through new combinations and increased homozygosity. Evolutionary biologists have explored this in both theory and experiments, and it helps explain why replicate founding events can lead to divergent outcomes even under similar selection pressures. The founding event "sets the table" by deciding which combinations are possible and how much variation selection can see.

Drift also interacts with recombination—the shuffling of genes during reproduction—in a way that matters for speed. In a tiny population, a favorable allele can be lost simply because it fails to be passed on. Recombination cannot rescue an allele that is gone, but it can change the fate of alleles that remain by breaking up unlucky associations. Still, in the earliest generations after a founding event, the population is often so small that the main story is not elegant shuffling but blunt sampling. The outcome can hinge on a few births, a few deaths, a few matings.

A historical anecdote from conservation biology captures the moral weight of these dynamics. When managers move a small number of individuals to establish a new population—a reintroduction, a translocation, a captive-breeding release—they are, knowingly or not, creating a founder event. The urge to "seed" a habitat quickly can conflict with the genetic need to carry as much diversity as possible. Too few founders can doom a population not because the habitat is unsuitable, but because the gene pool is too thin to buffer accidents, diseases, and future environmental changes. The early decades of many reintroduction efforts, from birds to mammals, have taught a sobering lesson: demographic success (getting numbers up) and genetic success (maintaining diversity) are not the same thing, and one does not guarantee the other.

The founder effect also carries a warning for how we interpret rapid evolution when we see it in the wild. Suppose an invasive species arrives in a new region and quickly develops a striking trait difference compared with its source population. The immediate assumption might be rapid adaptation to the new environment. Sometimes that is true. But invasions almost always involve bottlenecks: a few individuals arrive first, then spread. The invasive population may begin with odd allele frequencies, and those frequencies can shape traits directly. Selection then acts on top of that. Untangling what changed because of drift from what changed because of selection is one of the central detective stories of modern evolutionary genomics, and it matters for practical decisions. If a trait difference is mostly drift, predicting future change based on environmental pressures may be harder. If it is selection, managing the pressure may slow the change.

There is also a deeper philosophical point, uncomfortable in its simplicity. Natural selection is often portrayed as a force that produces fit organisms by weeding out the unfit. Drift reminds us that evolution is not always a meritocracy. Lineages can flourish because they were lucky early on. Alleles can become common because they were carried by a founder who happened to have many descendants. A population can become distinctive because a storm blew the wrong raft to the right shore. The living world is full of such accidents fossilized into genomes.

This does not make the world random in the everyday sense. Selection is real and often strong. But in small populations, the boundary between "strong" and "not strong enough" is easy to cross. An allele that would be reliably favored in a population of a million can be lost by chance in a population of a hundred. Once lost, it cannot help, no matter how useful it would have been.

The founder's gamble, then, is not only about losing options. It is also about becoming stuck with whatever options remain. A small population that has drifted into a particular genetic configuration can find itself constrained. Some evolutionary routes become inaccessible; others become easy. If the environment shifts, the population's fate may depend less on the direction of selection than on whether the necessary raw material survived the founding gate.

This is why small populations so often carry a sense of fragility. They are vulnerable not only because there are few bodies to spare, but because there are few genetic paths forward. A large population can afford to carry inefficient, redundant, even slightly costly variants that may someday become lifesaving. A small one is forced into austerity by chance alone, trimming its warehouse without consulting future needs.

And yet there is a stubborn persistence in life's arithmetic. Even in small populations, drift does not act alone. A rare beneficial variant, if it survives the early randomness, can rise quickly under strong selection. New mutations continue to appear. Occasionally, migration restores diversity. The gamble is never purely loss; it is risk distributed over generations. When a founding group succeeds, when a bottlenecked species rebounds, the resulting population often bears the genetic scars of its narrow past—and, sometimes, a few peculiar features that were never chosen for any purpose at all, but that now define what the population is.

2.3 Highways of Heredity

A population can be thought of as a conversation carried on in genes: who mates with whom, who survives, which variants spread. Migration adds a new voice mid-sentence. Sometimes it is a gift, delivering exactly the right phrase at exactly the right moment. Sometimes it is a heckle, drowning out a local dialect until everyone sounds the same.

The movement of individuals between populations—and the mixing of their genes through reproduction—is called *gene flow*. It sounds gentle, like a stream. In practice it can be as forceful as a floodgate. A few migrants per generation can be enough to alter evolutionary outcomes, especially when the receiving population is small. This is where rapid evolution becomes counterintuitive. Isolation feels like it should produce innovation, while mixing feels like it should produce stability. Often the opposite happens. Connection can speed adaptation by importing useful variants; it can also prevent adaptation by constantly replenishing the very alleles that local selection is trying to remove.

The earlier theme was that populations rarely wait for brand-new mutations; they rummage through what they already carry. Gene flow expands the rummaging space. The warehouse is no longer just what a population happened to inherit from its past. It becomes what its neighbors have tried and tested as well. A population facing a new stress can sometimes be rescued not by its own hidden diversity, nor by a lucky mutation, but by an immigrant.

There is a related idea with a practical ring: *genetic rescue*. Conservation biologists use it to describe what happens when introducing migrants into an inbred, small population increases fitness and survival. Part of this is simple arithmetic: migrants bring in alleles that were lost during bottlenecks, reducing inbreeding and masking harmful recessive variants. But genetic rescue can also be adaptive in a more direct sense. Migrants may carry alleles that help with disease resistance, climate tolerance, or fertility. When this happens, gene flow is not merely a demographic boost. It is an infusion of options.

The most famous conservation story along these lines involves Florida panthers. By the early 1990s, the remaining population was small and inbred, with signs of reduced fitness. Managers introduced individuals from a related population of pumas from Texas. Over time, genetic diversity increased and some health indicators improved. The debate around such interventions is never purely scientific—it touches identity, purity, and human responsibility—but as an evolutionary phenomenon it highlights a general point: when populations become isolated and small, the lack of gene flow can turn drift and inbreeding into a trap. A modest influx can reopen routes that chance had closed.

Yet gene flow is not automatically beneficial. The same migrants that bring new options can also disrupt local specialization. A mountain plant adapted to a short growing season can be swamped by pollen from valley plants that grow faster but cannot survive late frosts. A fish population adapted to a particular river's temperature and flow can be diluted by migrants from other streams. Even in a large population, if migration is constant and strong, local selection may never gain traction. This is sometimes called *migration load*: the burden of maladapted genes arriving from elsewhere.

To see why this can happen, it helps to make the logic concrete. Imagine a coastal population of an insect evolving resistance to a pesticide sprayed along shoreline farms. Resistance is helpful on the coast, but perhaps costly inland where the pesticide is absent: resistant individuals may grow more slowly or invest in a detox enzyme they do not need. Coastal selection pushes the resistance allele up;

inland selection pushes it down. If the two populations exchange migrants, the inland population supplies a steady stream of susceptibility alleles to the coast, and the coast supplies resistance alleles inland. The balance depends on numbers. If migration into the coast is heavy, the resistance allele may never become common enough to protect the population, even though it is favored there. The coast keeps receiving genetic "noise" from the inland environment.

This is not an exotic corner case; it is a common pattern in nature because environments are patchy. City blocks differ from parks, agricultural fields from hedgerows, shaded streams from sunlit ones. Gene flow stitches these patches together. Sometimes that stitching spreads a useful adaptation rapidly. Sometimes it prevents fine-tuning.

Human activity has changed the strength and direction of gene flow on a massive scale. Roads, canals, ships, and planes create new corridors. Habitat fragmentation creates new barriers. A species that once lived in a continuous forest now survives in isolated pockets separated by highways and suburbs. The consequence is not simply fewer individuals; it is fewer connections. In one place gene flow becomes too weak, allowing drift and inbreeding to rise. In another place gene flow becomes too strong, as invasive species hitch rides and mix formerly distinct populations. The modern world is a patchwork of newly opened and newly closed genetic highways.

Gene flow is not only about migration between populations; it also occurs *within* genomes. In sexually reproducing organisms, genes are reshuffled each generation by *recombination.* Homologous recombination happens when paired chromosomes exchange segments during meiosis, creating new combinations of alleles. This is a kind of hereditary traffic system. Without recombination, alleles would travel only in the packages they happened to be born into, and selection would have to choose between whole packages rather than individual items. With recombination, selection can assemble good combinations and break apart bad ones.

The importance of recombination becomes especially clear when many genes contribute to adaptation. If a population needs to adjust along multiple dimensions—say, heat tolerance, drought resistance, and timing of reproduction—then beneficial variants at different loci must end up together in the same individuals for selection to make full use of them. Recombination is the mechanism that allows this assembly, generation after generation. It can also prevent a beneficial allele from being trapped on a genetic background loaded with harmful variants. In that sense, recombination increases the efficiency of selection, helping populations respond faster to changing conditions.

Microbes complicate this story, because many of them do not rely on sex in the familiar sense. Bacteria reproduce by cloning themselves, which might seem to imply that recombination is irrelevant. Yet bacteria have another, more startling highway: *horizontal gene transfer* (HGT), the movement of genes between organisms that are not parent and offspring. Through mechanisms like transformation (picking up DNA from the environment), transduction (DNA moved by viruses), and conjugation (direct transfer via cell-to-cell contact, often mediated by plasmids), bacteria can acquire genetic material from unrelated lineages.

Horizontal gene transfer deserves its reputation as a shortcut. A bacterium does not need to wait for a long chain of mutations to build a new capability if that capability already exists elsewhere. It can, in effect, borrow a fully formed tool. This is particularly dramatic with antibiotic resistance. Resistance genes can move between bacterial species, often carried on plasmids that are built for mobility. In a hospital ward, where antibiotics create intense selection, the spread of resistance is not merely a matter of a resistant strain outcompeting a susceptible one. It can be a matter of resistance genes hopping into multiple strains and species, turning the genetic landscape into a network rather than a tree.

A historical vignette helps bring the microbial version of gene flow into focus. When antibiotics were first introduced widely in the 1940s and 1950s, they were hailed as wonder drugs. Resistance appeared quickly, and at first it was tempting to imagine it arising separately in each bacterial lineage as new mutations. Over time it became clear that some resistance traits were spreading through gene transfer, not just through descent. The metaphor of arms races remained useful, but the battleground changed. It was no longer simply a contest between two lineages evolving step by step. It became a contest in which bacteria could share weapons. The evolution of resistance, in other words, was partly social.

This also helps explain why antimicrobial use can have effects that seem disproportionate. Antibiotics not only select for resistant organisms; they can also change microbial ecology in ways that increase contact between strains, or increase the relative advantage of carrying mobile genetic elements. Some stresses can stimulate bacterial mechanisms involved in DNA uptake and transfer. The details vary with species and context, but the broader lesson is that selection and gene flow can reinforce each other. The result is rapid evolution not because mutation sped up dramatically, but because the traffic patterns of genes changed.

The same logic applies outside microbiology whenever gene flow moves adaptive variants across a landscape. One population discovers a solution by selection acting on its standing variation; migration exports that solution; recombination integrates it into new genetic backgrounds. When this works, adaptation can spread far faster than if every population had to solve the problem independently. This is one reason why insecticide resistance can appear “everywhere” in a region within a few years. It may not be dozens of independent inventions. It may be a combination of a few origins plus efficient movement.

But there is a trade-off, and it is not merely theoretical. If environments differ strongly, and if gene flow is high, populations may be prevented from specializing. The outcome is a kind of compromise phenotype: good enough everywhere, optimal nowhere. Whether that compromise is tolerable depends on how sharp the environmental differences are. A small mismatch might be harmless. A large mismatch can be fatal.

This tension can be framed as a balance between two rates: the rate at which selection increases locally favored alleles, and the rate at which migration introduces alleles favored elsewhere. If local selection is strong and migration is weak, local adaptation proceeds and populations diverge. If migration is strong relative to selection, populations homogenize. In the middle, something more dynamic can happen: stable differences maintained despite mixing, with steep genetic gradients across short distances. These gradients are called clines. They are nature's way of showing where gene flow and selection are in a tug-of-war.

Human decisions can tilt this balance abruptly. Consider salmon and other migratory fish in river networks. Dams and altered waterways can block gene flow between spawning populations, turning a once-connected set of lineages into isolated fragments. Each fragment becomes more vulnerable to drift and inbreeding, and its adaptive potential may shrink. Conversely, hatchery programs that move fish between rivers can increase gene flow, sometimes introducing alleles that reduce local fitness by disrupting adaptations to specific streams. Managers face an uncomfortable question: when does connection rescue, and when does it erase?

Even when gene flow is beneficial, it can carry a cost called *outbreed-*

ing depression, where crossing between populations reduces fitness because it breaks up co-adapted gene complexes or mismatches organisms to local conditions. This is less common than inbreeding depression in many conservation contexts, but it is real enough to warrant caution. In genetic terms, local populations can be finely tuned machines. Importing new parts may improve performance, or it may introduce compatibility problems.

The most vivid contemporary arena for gene flow's double edge is arguably agriculture and public health. Fields and hospitals are landscapes shaped by intense selection and by rapid movement. When a resistant variant arises, humans inadvertently help it travel: through trade, travel, water systems, and the movement of livestock and produce. This is gene flow in the broad sense, a network not only of genes but of bodies carrying genes. Efforts to slow resistance therefore cannot rely solely on reducing selection pressure (using fewer antibiotics or rotating pesticides), though that is crucial. They must also consider connectivity: which populations are mixing, which strains are being moved, which reservoirs of resistance are being tapped.

At the same time, the existence of gene flow can be an ally. In some cases, introducing susceptible individuals can slow the evolution of resistance by diluting resistant alleles, especially when resistance carries a cost in the absence of the drug. This is the logic behind certain refuge strategies in pest management, where areas without pesticide are maintained to preserve susceptible genotypes that can mate with resistant ones, reducing the fraction of resistant offspring. It is an attempt to use gene flow as a brake rather than a gas pedal. Such strategies are not foolproof, and their success depends on ecological details, but the principle is striking: the same mechanism that can spread adaptation can sometimes be harnessed to impede it.

The traffic metaphor is useful because it captures the moral neutrality of movement. Roads do not care whether they deliver medicine or contagion. Gene flow does not care whether it delivers rescue or erosion. It is a mechanism, and what it does depends on context: how different the environments are, how strong selection is, how large populations are, and how compatible the genetic backgrounds are.

In the end, rapid evolution is rarely a story of a single population alone. It is a story of connectedness: connected to its own hidden variants, connected to its past bottlenecks, connected to its neighbors through migrants, connected within its genome through recombination, and, for microbes, connected across species boundaries through horizontal transfer. When conditions shift, survival can depend on whether the right variant is present locally, whether it can arrive from elsewhere, and whether it can be integrated without being washed away by a tide of incoming genes. The outcomes we notice—a pest that suddenly shrugs off a pesticide, a small population that rebounds after a few migrants arrive, a local form that disappears into genetic sameness—are the surface ripples of those deeper currents of heredity.

Chapter 3

Invisible Armies

In the microscopic theater of war, the traditional rules of slow, generational inheritance do not apply. Bacteria treat their genetic code as open-source software, trading survival strategies across species lines faster than we can invent countermeasures. This chapter reveals how these invisible pathogens construct physical fortresses and engage in a relentless evolutionary arms race, forcing us to rethink our reliance on static chemical warfare.

3.1 The Blueprint Swappers

If evolution were only a family affair—traits passed from parent to offspring, generation after generation—bacteria would seem oddly constrained. Many of them divide every twenty minutes under good conditions, which sounds fast, until you remember what they face: antibiotics poured into wounds, lungs, and bloodstreams; disinfectants on hospital rails; heavy metals in wastewater; immune systems that learn and adapt. A microbe that relies solely on waiting for a lucky mutation and then patiently breeding it into the population is playing a dangerous game of timing.

Bacteria have a second option. They can acquire useful genes from their neighbours in the middle of their own lifetimes. Biologists call this *horizontal gene transfer*: genetic material moving sideways between contemporaries rather than vertically down a lineage. The effect is less like hereditary resemblance and more like swapping tools across a fence. An organism can wake up susceptible and go to sleep resistant, not because it "tried harder," but because it received a piece of DNA that already solves the problem.

The most vivid version of this swap involves *plasmids*. A plasmid is a small, usually circular, piece of DNA that lives inside a bacterial cell but outside the main chromosome. If the chromosome is the city's central archive, plasmids are portable thumb drives. They tend to carry genes that matter in emergencies: enzymes that break down toxins, pumps that eject antibiotics, proteins that help a cell stick to surfaces, and the molecular machinery that lets the plasmid copy itself and move into a new host.

Plasmids travel by a process called *conjugation*. Two bacterial cells come into contact; one extends a thin molecular bridge (often called a pilus) and transfers a copied strand of the plasmid DNA into the other. The recipient cell then completes the copy, closes the circle, and suddenly owns the same genetic upgrade. Crucially, this is not limited to identical clones. It can occur between quite different bacteria, even between species that would never "mate" in any ordinary sense. The bacterial world is full of such opportunistic exchanges, so that a gene's history is often not the history of the organism carrying it.

Plasmids are not the only route. Bacteria can also pick up naked DNA from their surroundings (*transformation*), a little like scavenging usable parts from a wreck. Others acquire genes via viruses that infect bacteria (*transduction*): a bacteriophage mispackages host DNA into its viral capsule and ferries it into the next cell it infects. But plasmids deserve their reputation as the great distributors because they are built for mobility. Many carry their own transfer apparatus. Some act as "hubs" that can recruit additional DNA segments and bring them along for the ride.

A detail that matters for human health is that resistance is often not a single gene but a *package*. Antibiotics attack bacteria in a small number of ways: they block cell-wall construction, jam protein synthesis, disrupt DNA replication, or sabotage key metabolic pathways. Bacteria counter with an equally small repertoire of strategies: they destroy the drug, modify the drug, alter the drug's target, prevent entry, or pump it back out. A plasmid can carry several of these at once. One DNA circle can encode an enzyme that cuts one antibiotic apart, a pump that ejects a second, and a protein that shields the target of a third. When such a plasmid moves, the recipient does not acquire a single trick; it acquires a toolkit.

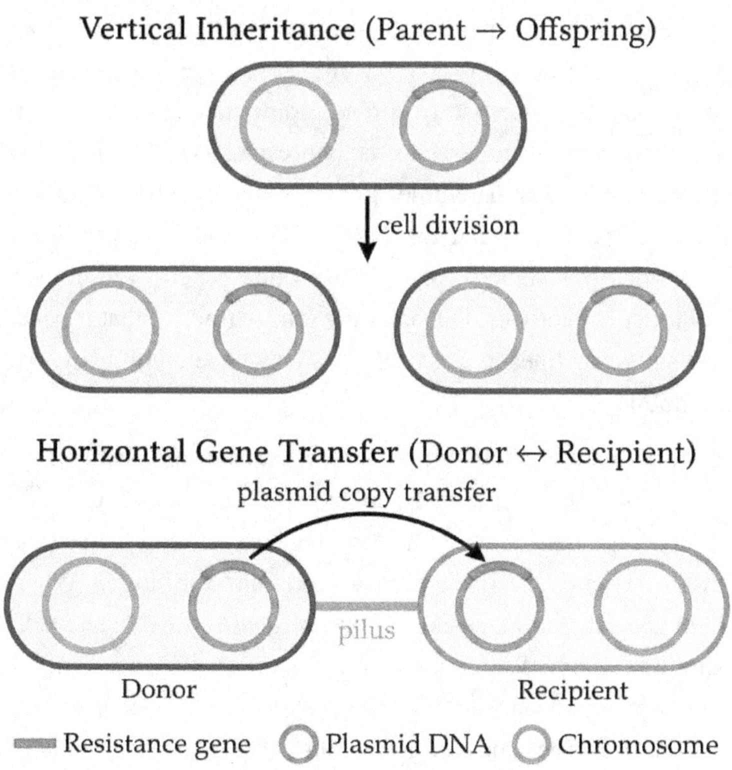

Once you picture plasmids as portable upgrades, a great deal of microbial behaviour becomes easier to understand. Antibiotic resistance does not need to arise anew in every patient, every ward, every city. It can be imported. A bacterium that has never "seen" a particular drug can nonetheless carry a gene shaped by long exposure elsewhere: perhaps in livestock treated with antibiotics, perhaps in a wastewater plant where residues and disinfectants mingle, perhaps in soil where natural antibiotic-producing organisms have been running chemical wars for millions of years. Human medicine becomes one arena in a much larger ecology of genes.

A historical turn makes the point concrete. In the late 1950s and early 1960s, doctors in Japan began reporting outbreaks of dysentery caused by *Shigella* that were resistant to multiple antibiotics. This was startling not merely because resistance existed—that was already becoming familiar—but because several resistances appeared together, and they seemed to spread rapidly across different strains. The explanation that emerged was that resistance determinants were being carried on transferable elements, later called R plasmids, that could move between bacteria. The clinical pattern—sudden, package-deal resistance—made sense once genes were allowed to travel independently of the bacterial "family tree." What looked like a new superbug lineage was, in part, a new gene combination riding on a mobile platform.

The mechanism is not magic; it is biochemistry and probability, amplified by selection. Imagine a mixed community of bacteria in the gut. Many are harmless commuters. One arrives carrying a plasmid with an enzyme that inactivates a particular antibiotic. When the patient takes that drug, susceptible bacteria are culled or slowed. The plasmid-bearing cells, by contrast, keep dividing. More importantly, every division produces more potential donors. Conjugation events that would otherwise be rare now occur in a population dominated by donors, and the plasmid can jump into bacteria that are better suited to cause disease. The drug does not merely favour resistant bacteria; it can favour a *gene's ability to spread.*

This can feel like a paradox: why would a bacterium help a competitor by donating a plasmid? The answer is that plasmids are not acts of generosity. They are replicating entities with their own evolutionary interests. Some impose a cost on their host: extra DNA to copy, extra proteins to express. Yet plasmids persist because they are good at copying and moving, and because they often carry "addictive" maintenance systems: molecular tricks that make it dangerous for the host cell to lose the plasmid once acquired. In a population under antibiotic pressure, a resistance plasmid can shift from a burden to a lifeline, and what benefits the plasmid and what benefits the host temporarily align.

Horizontal gene transfer also changes the meaning of *innovation.* In animals and plants, complex traits typically require many coordinated genetic changes and so evolve gradually. In bacteria, a multigene metabolic pathway can sometimes arrive as a block: a set of genes that together allow the use of a new nutrient or the neutralization of a toxin. A cell does not need to evolve the pathway step by step; it can import a working version. This is one reason bacterial genomes often look like patchworks. Some genes are ancient and conservative, essential to basic cellular function. Others are more like accessories, acquired and discarded depending on the cell's current neighbourhood.

To make this less abstract, consider what it takes to resist a β-lactam antibiotic such as penicillin. One route is to produce a β-lactamase enzyme that breaks the drug's key chemical ring. This is not a subtle tweak to an existing protein; it is a distinct function that can be gained by acquiring a gene that already encodes it. Once a plasmid provides that gene, the cell suddenly survives doses that would previously have been fatal. It is difficult to overstate what this means in clinical time. A patient's infection can shift from treatable to stubborn over the course of days, not because the bacterium divided enough times to "earn" a mutation, but because the right DNA happened to pass through the right encounter.

There is a temptation to interpret this as bacteria "cheating" Darwin. In reality, Darwinian logic still applies: variation, selection, and inheritance. Horizontal transfer simply expands what counts as variation and accelerates the mixing. Instead of waiting for a mutation to arise in the lineage you happen to be treating, you must contend with the possibility that the solution already exists somewhere in the local gene pool and can be imported. Selection then acts not only on bacterial clones but also on mobile genetic elements: plasmids, transposons (DNA segments that can hop within and between molecules), and integrons (genetic platforms that can capture and express gene cassettes). The microbial world is a marketplace of parts, and antibiotics function like a sudden spike in demand.

That marketplace has geography. A hospital is not just a building full of patients; it is a dense network of microbial habitats connected by hands, instruments, sinks, ventilators, drains, and antibiotics. In such a place, bacteria are exposed to intense selective pressures and frequent opportunities for contact. The very practices meant to control infection—broad-spectrum antibiotic use, heavy disinfection, invasive devices—can create conditions where plasmids carrying resistance are valuable, and where bacteria that would ordinarily live separate lives are brought into proximity. Horizontal gene transfer thrives on proximity. So does its darker consequence: resistance moving into organisms that are already well adapted to invade human tissue.

A second, more personal anecdote brings the phenomenon down to scale. When a urinary tract infection is treated with an antibiotic, the goal is simple: remove the pathogen. Yet the gut, skin, and urinary tract are not sterile battlegrounds containing only the enemy. They are crowded ecosystems. The drug becomes a selective filter that reshapes this ecosystem, favouring organisms that already carry resistance genes and encouraging the persistence of the plasmids themselves. Even if the immediate infection clears, the patient may temporarily carry a higher fraction of resistant bacteria. In other words, treatment can change the local gene pool, not only the local cast of species. What returns to the world from that patient—through contact, surfaces, wastewater—is not just bacteria, but a set of genetic options.

Horizontal transfer also forces humility about prediction. If resistance were purely a matter of stepwise mutation, one might estimate a probability and a timeline: how long until the right mutation appears? But if the relevant gene can be imported, the limiting factor is often social rather than genetic: how connected are the bacterial communities, and how strongly are we selecting for the mobile elements that travel between them? The "evolutionary clock" is replaced by a network problem.

Even so, there are barriers. Not every bacterium can accept every plasmid. Cells have defence systems that recognize and cut foreign DNA, analogous to immune systems at the molecular level. Plasmids, for their part, evolve ways to evade or suppress these defences. Compatibility matters too: a plasmid must replicate in its host, and its genes must be expressed without causing intolerable harm. These constraints mean horizontal transfer is not a free-for-all. It is more like trading between markets with customs checks, smugglers, and contraband. But the existence of barriers should not comfort us too much. Barriers create selective pressure for elements that can cross them, and the modern world provides repeated opportunities to try.

If all this makes bacterial evolution sound unnervingly agile, that is because it is. Yet there is a final twist worth holding onto: horizontal gene transfer is not only a villain in a clinical narrative. It is also part of how bacteria rapidly adapt in ways that benefit us. The same mechanisms that spread resistance can spread useful metabolic capacities in environmental microbes: the ability to degrade pollutants, to survive in harsh industrial conditions, to perform chemical transformations we harness in biotechnology. Nature's genetic swap-meets are morally indifferent. They become dangerous when our actions turn certain genes into tickets to survival and when our infrastructures bring the right partners into contact.

A patient with pneumonia does not experience plasmids; they experience breathlessness, fever, and the anxious waiting for a drug to work. A clinician experiences a lab report: resistant or susceptible, narrowing options. Somewhere beneath those human realities, an invisible exchange may have taken place: a loop of DNA copied, transferred, and closed, carrying a small red segment of information that changes what a microbe can endure. When that information moves sideways through a community, evolution stops looking like a slow staircase and starts looking like a sudden, crowded handoff.

3.2 Fortresses of Slime

A hospital bedrail looks like polished certainty: smooth metal, wiped down, gleaming under fluorescent lights. A catheter looks even more controlled: sterile packaging, a sealed pathway into the body, an object designed to be temporary and obedient. Yet some of the most stubborn infections begin not with a new, resistant mutant, or even with a newly acquired resistance plasmid, but with a different sort of adaptation: bacteria settling down, sticking to a surface, and building a communal shelter.

That shelter is a *biofilm*. The term sounds like a thin coating, and in a sense it is: a layer of microbial cells attached to a surface. But the crucial word is the one people tend to ignore: *film.* Films have structure. They can be thick or thin, patchy or continuous, porous or dense. They can channel flows, trap particles, and insulate what lies beneath. A biofilm is less like dust on a shelf and more like a living fabric, spun by bacteria from secreted molecules and held together by physics as much as by genetics.

The "slime" in popular descriptions is real, but misleadingly vague. Microbiologists often call it the extracellular matrix, meaning the material outside the cells that the cells themselves produce. It can contain sticky polysaccharides (long sugar chains), proteins that cross-link and strengthen the scaffold, lipids, and even strands of DNA released from cells and repurposed as structural fibre. Under a microscope, mature biofilms are often riddled with water channels and micro-caves, a kind of architecture that permits nutrients to circulate while keeping cells anchored and shielded.

The evolutionary advantage is immediate. A free-swimming bacterium in a fluid environment is exposed on all sides. An antibiotic molecule diffuses toward it, reaches its targets, and begins doing damage. A bacterium embedded in a matrix experiences a different world. Some antibiotics are slowed, trapped, or chemically modified before they reach the deeper layers. Immune cells struggle to grab and engulf what they cannot easily touch. Mechanical forces that would wash planktonic cells away can be resisted when cells are glued to each other and to the surface. The biofilm is an environment the bacteria build and then exploit.

The most unsettling feature is that biofilms allow survival without the kind of genetic resistance that shows up clearly in laboratory tests. Clinicians usually ask a simple question: is this bacterium resistant or susceptible? The answer comes from testing bacteria grown as a dispersed culture in a dish, where cells divide rapidly and antibiotics can bathe them evenly. Biofilm bacteria often behave differently. They can withstand concentrations of antibiotics far above what would kill the same strain in its free-living state. This is often called *tolerance*: the cells survive the drug exposure, but not because they carry a heritable mutation that defeats the drug's mechanism. They survive because their circumstances and physiology have changed.

That distinction matters because tolerance can be the quiet prelude to true resistance. If a biofilm keeps enough bacteria alive long enough, it extends the window in which a resistant mutant might arise or a resistance gene might arrive by transfer. A biofilm is a fortress that buys time. In evolutionary terms, time is opportunity.

A concrete, familiar example sits in many people's mouths. Dental plaque is a biofilm. You can feel it with your tongue: a fuzzy layer that accumulates even when you eat nothing exotic and do nothing dramatic. Plaque is not merely food residue; it is a microbial community attached to enamel and embedded in a matrix. The reason brushing works is not that it poisons bacteria, but that it physically disrupts the structure. Mouthwash can help, but the most reliable method is mechanical shear: break the film and you expose the inhabitants. The same logic scales up, with higher stakes, on a catheter, an artificial joint, or a heart valve.

The clinical stakes became hard to ignore in the late twentieth century. J. William Costerton and colleagues argued in the 1970s and 1980s that many persistent infections were biofilm infections: bacteria attached to surfaces, protected by matrix, and behaving differently from their laboratory counterparts. By 1999, Costerton, Stewart, and Greenberg were describing biofilms as a common cause of persistent infections in language that clinicians could not easily dismiss. It helped to name what doctors were already seeing: a patient who improves on antibiotics, then relapses; a culture that looks susceptible, yet the infection refuses to clear; an implant that becomes the centre of chronic inflammation until it is removed.

Removal is a brutal kind of proof. If an infection cannot be cured until the foreign surface is taken out, that surface is doing more than hosting bacteria. It is enabling a way of life.

Biofilms form through a sequence that is deceptively simple. First, individual cells attach to a surface. The attachment can be weak at first, mediated by physical forces and transient adhesins. Then they begin to produce matrix, strengthening the connection and recruiting other cells. As the community matures, it becomes chemically and physiologically diverse. Gradients form. Cells near the surface have more access to oxygen and nutrients; they may grow rapidly and experience stress from the outside world. Cells deeper down live in relative scarcity; they may grow slowly or not at all. Waste products build up in pockets. pH shifts. Oxygen disappears. The biofilm becomes a landscape with multiple niches only micrometres apart.

Those gradients matter because many antibiotics are most effective against actively dividing cells. A drug that blocks cell-wall synthesis, for example, hits hardest when bacteria are building new walls during division. If a large fraction of a biofilm's cells are slow-growing, they become less vulnerable, not because they have outsmarted the drug genetically, but because the drug's target is barely active. This is one reason the fortress metaphor is useful: fortresses do not only keep enemies out; they change the daily routines of those inside, allowing them to wait out a siege.

Within that waiting population, some cells enter an even more extreme state. They are often called *persisters*. A persister is not a genetically distinct strain; it is a phenotypic variant, a temporary physiological mode that makes the cell unusually tolerant to antibiotics. The word can be confusing because it sounds like an inherited trait. The key is that persisters *persist* through treatment and can later regrow, producing offspring that are just as susceptible as the original population when tested in the standard way. They are like sleepers, metabolically quiet, difficult to kill with weapons designed for busy factories.

Biofilms tend to enrich for persisters because biofilm life is full of stress signals, nutrient limitations, and local starvation. It is a setting that rewards flexibility: some cells work, some conserve, some hedge their bets. If antibiotics arrive, the hedgers become the founders of the rebound.

This is also where the simple story that "the slime blocks the drug" becomes too tidy. Diffusion barriers do exist. Some antibiotics penetrate poorly; some bind to matrix components; some are deactivated by enzymes concentrated near the biofilm's periphery. But in many systems, the main issue is not that the drug cannot physically reach the cells. It is that the cells it reaches are physiologically difficult to kill. A biofilm is both a material barrier and a way of reorganizing bacterial life into states that standard antibiotics are ill-suited to eliminate.

The fortress has another advantage: cooperation and division of labour. Bacteria are often portrayed as solitary, but many species communicate chemically, releasing small molecules that coordinate group behaviours. When enough cells are present, these signals accumulate and trigger changes in gene expression, including matrix production and virulence factors. The result is a community that can behave like a multicellular organism in miniature. Infections become less a matter of isolated invaders and more like an entrenched settlement.

Hospitals provide the perfect infrastructure for these settlements. Consider a ventilator tube. It is warm, moist, and repeatedly bathed in secretions. It offers a surface that bypasses the body's usual defences, and it sits where bacteria can access both nutrients and a route into vulnerable tissue. Biofilms can form on such devices quickly, and once established they can shed cells or clumps into the airway. Antibiotics may reduce the free-floating population, improving symptoms, while leaving the attached community intact. When treatment stops, the biofilm repopulates the area. The patient's story becomes a cycle of apparent victories and returns.

A similar dynamic plays out in chronic lung infections in cystic fibrosis, where *Pseudomonas aeruginosa* often forms biofilm-like aggregates in mucus. These infections are notoriously difficult to eradicate. Part of the difficulty is classical resistance, which can evolve under repeated antibiotic exposure. But part of it is the fortress strategy: a protected, heterogeneous community in which some cells are shielded physically and others are shielded by their slow, altered physiology. The lung becomes a landscape where bacteria can carve out protected niches, and the immune response, though fierce, struggles to reach and clear them completely.

Biofilms also help explain why certain infections cling to particular places. Endocarditis, an infection of heart valves, often involves bacterial attachment to damaged tissue and the formation of vegetations—dense masses of bacteria, platelets, and host proteins. The structure is not identical to a textbook biofilm on plastic, but it shares the same logic: attachment, embedding, and protection. Likewise, infections of artificial joints can smoulder for months. Bacteria on the surface can be sparse enough to evade detection, yet stubborn enough to survive. The immune system cannot easily clear a foreign surface, and antibiotics may not reach effective killing dynamics within the biofilm's protected microenvironments.

One of the most frustrating clinical consequences is diagnostic. A biofilm can be present even when cultures are negative or inconsistent. If bacteria are attached to an implant and living slowly, they may not shed many cells into the sampled fluid. When they do shed, the cells may be in a different physiological state once they reach the laboratory. The fortress makes the enemy hard to count.

Biofilms are not just a medical issue. They are an evolutionary strategy that makes sense in streams, soils, and plumbing systems as well. In nature, living attached to a surface can be safer than drifting. Nutrients often concentrate on surfaces. A rock in a stream develops a microbial mat; plant roots are coated with bacterial communities; the inside of a household sink drain is a richly textured biofilm ecosystem. Antibiotic tolerance is not the "purpose" of these structures. It is a consequence of the protection and heterogeneity that attached life provides. When humans apply antibiotics, we collide with a mode of existence that has evolved for many reasons, but happens to frustrate our drugs.

This also reframes a common misconception: that bacteria become hard to kill only by changing their genes. Genetic resistance is powerful and permanent, and it deserves the attention it receives. But biofilms show that evolution can act on behaviour and collective structure. Natural selection can favour bacteria that are better at attaching, better at producing matrix, better at coordinating as a group, better at enduring long periods of stress. Those are traits. They are heritable. They can evolve rapidly. They are not as easily spotted as a resistance enzyme, but they can be just as clinically decisive.

The distinction between resistance and tolerance can be felt in the difference between *defeating* a drug and *outlasting* it. Resistance is a lock changed so the key no longer fits. Tolerance is staying behind a door until the person with the key gives up and leaves. Antibiotic courses are finite. Hospital admissions are finite. Even immune responses wax and wane. A strategy that reliably buys survival through those windows can be favoured, even if the bacteria remain genetically vulnerable in principle.

The most hopeful part of this story is that physical strategies invite physical counter-strategies. If plaque can be disrupted by brushing, then medical biofilms can sometimes be countered by designing surfaces that resist attachment, by using coatings that interfere with matrix formation, by employing enzymes that degrade matrix components, or by applying carefully timed combinations of antibiotics that target both active and dormant subpopulations. None of these is a perfect fix, and each invites its own evolutionary replies. But biofilms remind us that antimicrobial therapy is not only chemistry. It is also engineering, ecology, and an uncomfortable kind of architecture.

A final image helps: imagine antibiotics as rain falling on a city. In an open field, rain reaches everything quickly. In a dense, layered city, some streets flood, some remain dry, and some residents hide indoors until the storm passes. If the storm comes often enough, the city may even expand its roofs and drains. Biofilms are bacteria learning to live in cities of their own making, and the matrix is their building material. Once bacteria become builders as well as breeders, survival is no longer just about which genes they carry, but about the worlds those genes allow them to construct.

3.3 The Infinite Arms Race

A new antibiotic tends to arrive with a familiar burst of confidence. There is a press release, a set of clinical trial curves, a sense that medicine has regained its footing. For a while it often has. Patients recover. Surgeons operate with less dread. Chemotherapy becomes safer. Then the reports begin to change tone. A laboratory notices a strain that "should" be susceptible but is not. A hospital infection-control team sees a cluster. A clinician reaches for a second-line drug sooner than expected. The pattern is not a surprise anymore, but it remains jarring: a chemical that looked like a decisive invention becomes, in a few years or a few decades, a temporary advantage.

The reason is not that bacteria are unusually clever. It is that antibiotics do not create an enemy from scratch; they select from an enemy that has been practising for a very long time. Many antibiotics are derived from compounds made by microbes in soil and water, where bacteria and fungi have fought chemical wars for hundreds of millions of years. If a microbe in the ground makes a molecule that harms its competitors, those competitors face strong pressure to evolve defenses. Some defenses involve altering targets so the antibiotic can no longer bind. Some involve enzymes that chew the molecule up. Others involve pumps that eject it. The genes for these defenses accumulate in microbial communities, often doing double duty: the same pump that gets rid of a natural toxin may incidentally remove a human-made drug. In that sense, resistance is not a modern invention. Modern medicine mainly changes the scale, the speed, and the direction in which old capabilities move.

The scale is easy to forget. Antibiotics are not used only in intensive care units. They are used in outpatient clinics, dentistry, agriculture, and animal husbandry; they enter rivers through wastewater; they seep into soils through manure and runoff. Even when concentrations are low, they can provide a steady nudge, favouring bacteria that carry resistance genes and, crucially, favouring mobile genetic elements that can spread those genes. Selection does not require a dramatic overdose. It requires a consistent difference in survival and reproduction, repeated across enormous populations. Bacteria provide the enormous populations for free.

The speed comes from two accelerants already on the table: the ability of bacteria to swap genes and the ability of bacteria to hunker down in biofilms and survive incomplete assaults. When a drug is deployed, it is not facing a single lineage evolving step by step. It is facing a network of genes that can move laterally and a set of microbial lifestyles that can outlast the treatment window. Those are not details; they are the engine of the modern antibiotic crisis.

The direction is perhaps the most unsettling part. An antibiotic is designed to attack a pathogen, but selection pressure rarely stays confined to the pathogen. A drug taken for pneumonia does not only touch the bacteria in the lungs; it touches the gut microbiome, the skin, the mouth. These communities are usually not dangerous, but they are vast reservoirs of genes. If resistance genes rise in frequency there, they become more available to move into pathogenic species later. The antibiotic has, in effect, trained and stocked the microbial world around the patient, not just battled the infection at hand.

This is why the arms race metaphor is accurate but incomplete. Traditional arms races involve two sides designing weapons against each other, with each innovation answered by a counter-innovation. With antibiotics, humans introduce a new chemical, and bacteria respond. But bacteria are not responding only as individual pathogens in the patient. They respond as an ecology: commensals, environmental bacteria, opportunists on hospital surfaces, microbes in drains and soils. Evolution is distributed. The countermeasure may evolve far from the bedside and arrive later by gene transfer.

A classic historical example is methicillin. Penicillin resistance, often mediated by β-lactamase enzymes that destroy the drug, became a major clinical problem within a few years of widespread penicillin use. Methicillin was introduced as a workaround: a β-lactam antibiotic designed to resist those enzymes. In 1961, just after methicillin entered clinical use, methicillin-resistant *Staphylococcus aureus* (MRSA) was reported. The key adaptation was not another β-lactamase but a modified target: an alternative penicillin-binding protein (PBP2a) that keeps building the cell wall even when the drug is present. The lesson is not that methicillin was poorly designed; it was that bacteria had multiple routes to the same outcome. When one door is blocked, selection pushes harder on another.

This multiplicity is the microbial advantage. A single antibiotic class may be countered by several distinct mechanisms, and bacteria can combine them. They can reduce drug entry by altering membrane porins. They can pump drugs out using efflux systems. They can alter the binding site so the drug no longer recognizes its target. They can chemically modify or degrade the drug with enzymes. They can bypass the targeted pathway altogether, using an alternative metabolic route. If evolution were forced to find a single narrow solution, it would be slower and more fragile. Instead, bacteria search a broad space of workable hacks.

Mobile genetic elements accelerate that search by bundling solutions together. A plasmid can carry a resistance enzyme, an efflux pump, and a regulatory gene that tunes expression under stress. Transposons can hop these modules into new contexts. Integrons can capture gene cassettes and express them under a shared promoter, turning a bacterium into a kind of plug-in platform. Under heavy antibiotic pressure, bacteria that acquire such elements gain immediate survival; the elements themselves gain a proliferating host. The result is not just resistance, but *multi-drug resistance*, where a single acquisition can reduce the effectiveness of several unrelated drugs.

Even when antibiotic use is reduced, resistance does not necessarily melt away. There is a comforting story in which resistance carries a fitness cost and therefore disappears when the drug is removed, like armour abandoned when peace returns. Sometimes that happens. Often it does not. Bacteria can acquire compensatory mutations that reduce the cost while maintaining the resistance function. Resistance genes may be linked to other advantageous genes on the same plasmid, so that selection for one trait maintains the whole package. Co-selection is particularly insidious: exposure to heavy metals or disinfectants can maintain plasmids that also carry antibiotic resistance genes, because the entire element is selected as a unit. In that way, a hospital's cleaning regimen and plumbing chemistry can indirectly shape antibiotic resistance patterns.

The human side of the arms race is constrained in ways bacteria are not. Developing an antibiotic is expensive, slow, and risky. The most obvious targets have already been exploited, and bacteria resemble us at the molecular level more than is convenient: many essential bacterial processes have counterparts in human cells, so drugs that disrupt bacteria often threaten human tissues as well. A truly novel antibiotic needs to be potent against pathogens, safe for patients, stable in the body, manufacturable at scale, and economically viable. And the moment it is introduced, stewardship programmes urge restraint, because heavy use will speed resistance. The paradox is that society wants new antibiotics urgently, but also wants them used sparingly, which reduces the return on investment. Evolution does not care about business models, but our ability to respond does.

This is why "stronger" drugs do not solve the problem in the long run. A stronger drug can apply stronger selection. When an antibiotic is used widely, it does not merely kill bacteria; it acts as a filter, letting through any cell that happens to carry a countermeasure. The sharper the filter, the more strongly it enriches the survivors. In evolutionary terms, antibiotics are unusually intense selection pressures: a small genetic change can mean the difference between death and uninhibited growth. In such conditions, rare variants become common with astonishing speed.

The most vivid illustration may be the way resistance can appear repeatedly and independently. Different hospitals, different countries, different bacterial lineages can converge on the same solution because the solution is available in the broader gene pool. If a particular enzyme can degrade a drug, or a particular mutation can alter a target, it may emerge multiple times. When that solution is carried on a plasmid, it can also leap between contexts. Evolution becomes both parallel (the same trick reinvented) and contagious (the same trick transmitted).

Against this background, bacteriophages—viruses that infect bacteria—look less like a footnote and more like a different philosophy of medicine. A conventional antibiotic is a static molecule. It does not adapt. If bacteria evolve resistance, the molecule remains the same and its usefulness declines. A phage is a replicating, evolving entity. It attaches to a bacterium, injects its genetic material, and uses the cell as a factory to make more phages, often killing the host in the process. As long as the phage population can keep pace with bacterial defenses, it can, in principle, remain effective.

Phage therapy is not new. It was explored in the early twentieth century, before antibiotics became dominant. Its history is tangled with scientific uncertainty, uneven manufacturing, and geopolitical accident; it persisted more continuously in parts of Eastern Europe and the former Soviet Union than in the West. The idea has returned with renewed urgency as resistance rises. The appeal is easy to grasp: fight an evolving enemy with an evolving weapon.

But phage therapy does not escape the arms race. It relocates it. Bacteria can become resistant to phages by changing the surface receptors that phages use to attach. They can deploy intracellular defenses that cut up phage DNA. They can use restriction-modification systems, which mark their own DNA with chemical tags and slice untagged foreign DNA. Some bacteria carry CRISPR-Cas systems—a kind of adaptive immune memory in which snippets of past invaders are stored in the bacterial genome and used to recognize and destroy matching sequences in future attacks. Phages, in turn, evolve countermeasures: altered binding proteins to recognize new receptors, proteins that inhibit bacterial defenses, and strategies that minimize the exposure of vulnerable genetic material.

CRISPR has acquired a second, more public life as a laboratory tool for gene editing, but in nature it is part of this conflict. It can block incoming plasmids and phages, shaping what a bacterium can acquire horizontally. That creates an evolutionary trade-off. A bacterium with strong defenses may be safer from viral attack, but it may also be less able to pick up beneficial genes such as antibiotic resistance determinants when those are needed. Conversely, a bacterium open to gene acquisition may be more vulnerable to genetic parasites. There is no perfect posture. The balance shifts with ecological circumstances: in a phage-rich environment, defense is valuable; under antibiotic pressure, openness to incoming resistance genes can be lifesaving. The result is a restless evolutionary landscape in which both immunity and exchange are continually renegotiated.

Phages also interact with biofilms in ways that underline the complexity of microbial warfare. A biofilm's matrix can impede phage movement, just as it can impede antibiotics, and bacteria in deeper layers may be metabolically inactive, offering fewer resources for phage replication. Yet some phages carry enzymes that degrade biofilm matrix components, helping them penetrate. In some cases, phages can act like self-amplifying agents within a biofilm: once they reach a pocket of susceptible cells, they replicate locally and spread. The outcome depends on timing, architecture, and the specific pairing of phage and host. It is not a simple replacement for antibiotics, but it is a reminder that living weapons have tactics unavailable to static molecules.

The arms race has a moral dimension that is easy to miss. When a patient receives antibiotics, they are not choosing to participate in evolutionary biology; they are choosing to breathe, to avoid sepsis, to survive surgery. Antibiotics remain among the most life-saving inventions in medical history. The tragedy is that the benefits are shared and the costs are shared, but in different ways. The benefit is immediate and personal. The cost is delayed and collective: increased frequency of resistance genes circulating in communities and environments, reduced effectiveness of future treatments, a narrower safety margin for modern medicine.

That collective cost becomes visible in the hospital outbreaks that make headlines: carbapenem-resistant Enterobacteriaceae, vancomycin-resistant enterococci, multi-drug resistant *Acinetobacter*. The names sound technical, but the human reality is simple. When last-resort drugs fail, clinicians fall back on older, more toxic treatments, on combinations with uncertain benefit, or on supportive care alone. Routine procedures become risky. Neonatal wards and intensive care units become places where evolution has immediate consequences.

Evolution's final insult is that it is not obliged to be fair. Resistance does not necessarily arise where antibiotic use is most justified. It can be amplified where use is excessive or poorly controlled, then travel. Genes move more easily than policies. A plasmid does not need a passport. A resistant strain can hitchhike on a human traveller, a food supply chain, a hospital transfer, a contaminated surface. The microbial world is wired for dispersal.

Yet it would be a mistake to see this as a story of inevitable defeat. Arms races do not always end with one side exhausted; they often end with shifts in strategy. Part of that shift is social and political: stewardship, surveillance, infection control, sanitation, and the careful use of antibiotics in agriculture. Part of it is scientific: drugs that disarm pathogens rather than kill them outright, reducing selection pressure; combinations that make resistance more costly; therapies that disrupt biofilms; and, in selected contexts, phages and other biological agents that can respond dynamically.

What remains stubbornly true is the central asymmetry: humans innovate in bursts, constrained by regulation, economics, and biology, while bacteria innovate continuously, constrained mainly by physics and the ruthless arithmetic of survival. Every antibiotic prescription is a small experiment in natural selection. Every hospital surface is a potential meeting place for genes. Every chronic infection is an ecosystem under siege. The microbes do not need to plan. They only need to vary, persist, and spread, and they have been doing that since long before the first human named a disease.

Chapter 4

Concrete Jungles

Cities are often viewed as biological deserts, yet they function as high-velocity evolutionary accelerators. By replacing forest floors with concrete and trapping solar heat in asphalt, we have inadvertently built a massive global experiment in natural selection. This chapter uncovers how the unique stressors of the metropolis-heat, smooth surfaces, and subterranean isolation-are reshaping the genetics of wildlife in real time.

4.1 The Furnace of Asphalt

A city changes the weather without asking permission. On a clear summer evening, when the countryside begins to cool, the streets keep radiating heat as if they were still in full sun. Brick, asphalt, tar roofs, and dark masonry behave like a giant storage heater: they soak up solar energy by day and release it slowly through the night. Add the waste heat of engines, air conditioners, and industrial vents, and the result is a familiar experience that barely needs a scientific name. The *urban heat island* effect describes the tendency of built-up areas to run hotter than nearby rural ones, especially after sunset, when vegetation and moist soils would otherwise cool the air through evaporation.

For people, this is discomfort and expense: restless nights, overworked cooling systems, elevated risk during heat waves. For small animals and plants pinned to the surfaces of the city, it is something sharper. Heat is not only a nuisance; it is a filter, and it is often a lethal one.

Thermal biology has an uncompromising quality. Many organisms can endure hunger, crowding, even toxins for some time, but temperature pushes directly on the chemistry of life. Enzymes lose their shape; membranes become too fluid; nerves fire erratically; muscles fail to contract properly. The problem is not that an organism prefers a certain temperature. The problem is that, beyond a boundary, the machinery stops working. Biologists often summarize that boundary with measures such as the *critical thermal maximum* (CTmax): the temperature at which an animal loses coordinated function and can no longer right itself, walk, cling, or escape. CTmax is not "the hottest day it can tolerate" in an everyday sense. It is a performance cliff, and cliffs matter most in a world full of predators, traffic, and hard edges.

Cities create more opportunities to fall off that cliff. Not only are average temperatures higher, but the distribution changes in a way that is especially cruel to small-bodied creatures. Sunlit pavement can reach temperatures that would seem absurd if reported as air temperature, and those extremes are exactly where many insects, spiders, and small lizards live their lives: along the ground, on walls, on the surfaces of leaves growing from thin soil in planters, and in the narrow gap between a building and the sidewalk. In the countryside, a patch of shade under a shrub may be enough to stay within safe limits. In the city, shade can be scarce or short-lived, and the "cool" refuge may still be warmer than the rural daytime air.

The evolutionary question is deceptively simple: when heat becomes more common and more intense, do populations adapt? The broader question is harder: *how quickly* can they do so, and what do they have to give up in return?

Natural selection needs three ingredients: variation (individuals differ), heritability (some of those differences are passed to offspring), and differential success (some variants survive and reproduce better than others). Urban heat islands supply the third ingredient with brutal efficiency. A single afternoon of extreme surface temperatures can act like a sieve. Those that cope survive to breed. Those that cannot, do not. When that pattern repeats across many summers, the city becomes an evolutionary furnace.

A compelling reason cities are scientifically useful is that they create "replicated" environments. One city is not an experiment; it is messy, with idiosyncratic history and ecology. But many cities share a set of physical features: dark surfaces, reduced tree cover, altered wind patterns, and heat released by human activity. If populations repeatedly evolve greater heat tolerance in independent cities, that repeatability is itself evidence that temperature is a driver rather than an incidental detail.

One of the clearest stories comes from an animal so small and so ordinary that most people never notice it: the acorn ant. Acorn ants live in the hollowed shells of fallen acorns and other small cavities, forming colonies that can be tucked into tiny spaces across forests and parks. Their homes are portable; their world is fine-grained. That makes them excellent witnesses to microclimates, and microclimates are where heat islands do much of their work.

In the Cleveland area, Sarah E. Diamond and colleagues (2017) compared acorn ant populations collected along an urban-to-rural gradient. The elegant part of the work was not simply measuring ants in the field, because field measurements always leave an obvious doubt: perhaps city ants behave differently, choose different nests, or experience different food conditions. Instead, they used a "common garden" approach. Ants from different sites were reared under the same laboratory conditions, so that persistent differences in thermal tolerance were less likely to be mere short-term acclimation. The pattern was striking. Ants originating from warmer urban sites tolerated higher temperatures than ants from cooler rural sites, while also showing reduced tolerance to cold. In other words, city life reshaped their thermal profile in a direction that fits the selective pressure.

That last detail matters because adaptation is often a trade-off. If the thermal window were a stretchable rubber band, one might hope it could expand in both directions: better heat tolerance *and* better cold tolerance. But physiology is not always so generous. Proteins that remain stable at higher temperatures may be less efficient at low temperatures; membranes optimized to remain intact during heat may become too rigid in the cold. There are also energetic costs: producing heat-shock proteins, repairing damage, maintaining ion gradients under stress. A city may reward an organism for spending more of its budget on heat defense, but that budget must come from somewhere.

Urban heat islands offer a particularly vivid way to think about how trade-offs become real. In many temperate cities, winters still arrive. The city may be warmer than surrounding rural areas, but it is not tropical. Yet an animal with reduced cold tolerance can still thrive if it is sheltered by the same infrastructure that created the heat: building foundations, subterranean spaces, heated basements, insulated walls, and human refuse that provides steady food. The city does not merely raise temperature; it rearranges the whole package of seasonal risks.

The pace is equally sobering. The warming associated with urbanization in many regions has been building over roughly a century. For humans, that is long enough to become "how it has always been." For an ant, a fly, or a mosquito, it can be only a handful of generations. Acorn ant colonies do not turn over every week, but their generation time is still short enough that evolutionary change can occur on a historical timescale. When we say "rapid evolution," it is easy to imagine something dramatic: new colors, new shapes. The first thing evolution often changes is more subtle and more important: thresholds, tolerances, and the ability to keep functioning when the environment tries to push you out of the viable range.

One reason heat is such a powerful selective force is that it makes failure immediate. Many hazards are probabilistic. A poor diet increases disease risk; a mild toxin reduces fertility; a slightly mismatched camouflage pattern raises predation risk. Heat can do all of those things too, but it also has a simple, non-negotiable option: death within minutes. For small-bodied ectotherms (animals whose body temperature is strongly shaped by the environment), the margin between safe and lethal can be narrow. A few degrees can separate "able to run" from "unable to move," and in cities those few degrees are routine.

There is an intuitive temptation to treat thermal tolerance as a single trait, as if a species has one number on a thermostat. In reality it is a suite: upper limits, lower limits, preferred temperatures, the slope of performance decline, and the ability to *acclimate* (to adjust physiology within a lifetime). Acclimation can be thought of as flexibility: if the weather warms for a week, can the organism shift its biochemistry to cope? Evolution can act on that flexibility too. Some lineages may thrive in cities not because their baseline CTmax is high, but because their plasticity is large enough to track the city's erratic heat.

That distinction becomes crucial when the environment changes faster than bodies can track. A recent comparative study by Campbell-Staton and colleagues (2024) synthesized evidence across diverse groups and found a broadly consistent pattern: urban populations tend to have higher heat tolerance and reduced cold tolerance relative to nearby non-urban populations. Yet the same work emphasized a tension that should make us cautious and curious. In many cases, physiological change appears to lag behind the magnitude of urban warming. The city turns up the heat; populations respond; but not always enough, not always in time. Adaptation is real, but it is not magic, and it is not guaranteed.

Why might response lag? One obvious reason is genetic limitation: perhaps there is not enough heritable variation in thermal traits for selection to act on. Another is gene flow: if city populations are constantly mixing with rural populations, alleles favored in the city may be diluted. Another is the nature of heat exposure. Extreme temperatures may occur as short spikes; selection may act strongly during rare events but weakly in average conditions. A final reason is ecological buffering. Cities contain a patchwork of microhabitats: shaded courtyards, irrigated lawns, air-conditioned exhausts, tree-lined streets, underground drains. If an organism can behaviorally avoid the hottest places, selection on physiology may be weaker. But that buffering is unevenly available, and it can disappear abruptly. A tree comes down; a parking lot is resurfaced; a drought limits irrigation; a heat wave extends into the night. A population that has relied on behavioral refuge can find itself suddenly exposed.

Heat is also an evolutionary force with a moral twist: it does not act equally on all organisms. Humans, with clothing and shelter and air conditioning, often experience urban heat as an inconvenience punctuated by dangerous episodes for the vulnerable. For small animals, the baseline can already be close to the edge. This asymmetry is one reason heat islands are more than a curiosity. They are a mechanism by which the built environment quietly sorts which species live alongside us.

A historical anecdote makes the point. During the great heat wave that struck Chicago in July 1995, hundreds of people died, many of them elderly and socially isolated, in neighborhoods where nighttime temperatures remained high. Public discussion focused *(rightly)* on social vulnerability and infrastructure. But the physical mechanism was the same one faced by a beetle under a streetlamp: the failure to cool at night. Nighttime relief is one of the most important protections against heat stress. Cities blunt that relief. For humans, the consequences are concentrated in particular groups; for many small ectotherms, the consequences are spread broadly, and they play out as selection, generation after generation.

Thermal evolution in cities is not confined to insects. In Puerto Rico, the crested anole lizard has become a symbol of urban adaptation for reasons that include its biomechanics, but it also lives in a thermal landscape shaped by concrete and sun. Urban anoles experience higher operative temperatures (the temperatures their bodies are pushed toward by the environment), and urban individuals often show higher heat tolerance than forest counterparts. In work on urban heat physiology, Campbell-Staton and collaborators (2020) reported signatures consistent with repeated selection on heat tolerance in city populations, and linked variation in thermal plasticity to genetic differences, including a notable variant in a gene involved in protein synthesis (RARS). The details of molecular biology can be left for specialists, but the logic is general: if city heat repeatedly rewards certain physiological responses, selection can find genetic routes to produce them.

It is worth pausing on what "genetic routes" means in practice. Evolution does not necessarily invent new genes. More often it tweaks regulation: when a gene is turned on, in which tissues, and how strongly. Thermal tolerance often involves networks of stress-response genes, membrane remodeling pathways, and metabolic adjustments. A change that slightly increases the stability of proteins at high temperature may help in the city while slightly reducing efficiency at low temperatures. A shift that increases the capacity to produce protective chaperone proteins may aid survival during heat spikes while costing energy during normal days. The city does not need to create a new trait from scratch; it can favor a different balance in the existing toolkit.

The urban heat island also forces a reconsideration of what "local adaptation" looks like. We tend to imagine local adaptation as a mountain population diverging from a valley population, separated by distance. Cities create sharp gradients over short distances. A park a few kilometers from downtown can be measurably cooler than a heavily paved district. That means selection can differ over distances an animal could traverse in a day, at least in principle. Whether it does depends on barriers, behavior, and life history. For organisms that disperse poorly or live in small, stable colonies, a city can contain multiple thermal worlds separated by roads, buildings, and habitat gaps. This is one reason urban evolution can be so rapid and so patchy: selection is strong, and movement is not always easy.

There is another subtlety. Heat islands do not simply raise temperatures uniformly; they change *variability*. The city can produce hotter peaks and warmer nights, flattening the daily cycle in ways that can be physiologically challenging. Many organisms rely on cool nights to repair heat damage. A day that is tolerable in isolation can become lethal if the night offers no recovery. In that sense, the city's most consequential heat may occur after dark, when the rural world is resting and the city is still glowing. A thermal tolerance measured under laboratory ramps is useful, but what matters in nature is how performance and recovery play out across real cycles of exposure.

To make this more concrete, imagine two ant colonies identical in every way except for a slight difference in CTmax, perhaps just a degree or two. In a rural setting where the hottest microhabitats are rare and nights are cool, that difference may barely matter. In a city, where heat spikes are common and nights stay warm, the colony with the higher CTmax may keep foragers active longer, gather more food, and avoid the cascading failures that come from repeated stress. Over a summer, that advantage can translate into more queens produced or higher colony survival. Over decades, it can translate into a shift in the population.

The simplest way to visualize such a shift is to plot CTmax for urban and rural populations across a temperature gradient. The values below are illustrative, designed to capture the qualitative pattern reported for acorn ants: urban-origin populations exhibit higher CTmax than rural-origin populations, consistent with adaptation to warmer environments.

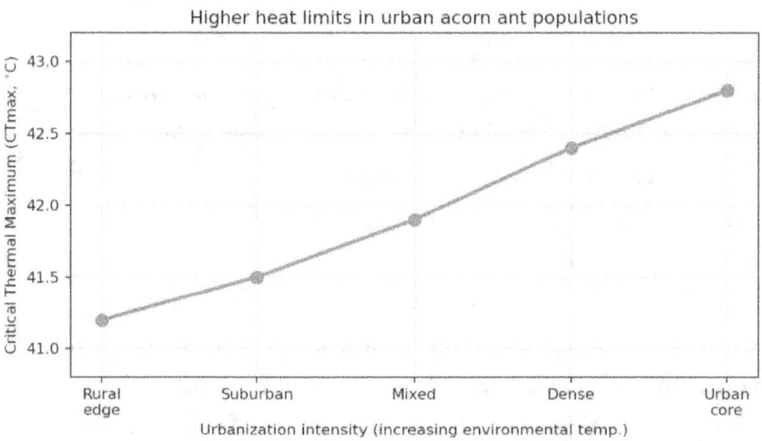

The figure invites an obvious question: if urban populations gain heat tolerance, why do they not simply become "better" across the board? Why does cold tolerance often decline? Partly because selection is not a committee dedicated to general improvement. It is a process that rewards whatever works *here and now*. In a city, surviving heat may be the difference between reproducing and not reproducing, while surviving cold may be buffered by protected microhabitats or by the fact that the urban winter is milder than the rural one. When the pressure on cold tolerance relaxes, variants that would have been eliminated in a colder setting can persist, and variants that enhance heat tolerance at the cost of cold performance can spread.

There is also a deeper lesson: adaptation can narrow as well as shift. A population can become specialized to the thermal regime it experiences most. That specialization may be advantageous in a stable environment, but it can carry vulnerability if conditions change again. A city that warms further, or that becomes more variable with climate change, may push even heat-adapted populations toward their limits. Conversely, a city that increases tree cover and green infrastructure may cool enough that the old trade-offs are no longer optimal. Evolution does not read urban planning documents; it responds to what the environment becomes.

Some of the most interesting evolutionary stories in cities are not about spectacular new features but about these quiet shifts in limits and balances. They matter because they determine which organisms can live in the thin strip of habitat we inadvertently provide. A crack in a sidewalk, a planted median, a row of street trees, a rooftop garden: these are not simply decorative elements. They are thermal habitats. When we pave a lot, we do not just remove plants; we create a heat landscape that selects for certain physiologies. When we plant trees, we do not just add shade; we change the selective filter itself.

The city, then, is not merely a place where nature persists despite us. It is a place where our materials and our habits write new rules of survival. In that furnace of asphalt, evolution does what it always does: it sorts among small differences, turning microclimates into life-and-death exams, and leaving, in the bodies of ants and lizards and countless overlooked species, a record of the temperatures we have built.

4.2 Scaling the Glass Cliffs

A tree trunk is an honest surface. It is rough, furrowed, and irregular, offering a thousand tiny ledges for claws and skin to catch. Even a smooth-barked sapling carries its own kind of grip: microscopic texture, patches of moss, a seam where bark plates meet. City architecture is a different planet. Painted concrete, glazed tile, aluminum, glass, and plastic are broad, hard, and often vertical. They erase the small-scale topography that many climbing animals have relied on for millions of years.

To a lizard, that difference is not aesthetic; it is physics. A slip on bark might end with a scramble and a short fall. A slip on a plastered wall can end with a long drop into open air, onto a sidewalk, or into a street. The city offers abundant perches, but they are perches that punish the wrong design.

The Puerto Rican crested anole (*Anolis cristatellus*) has become a small celebrity in urban evolutionary biology for the simple reason that it is willing to live where people do, and it shows its adjustments in plain view. On islands and in tropical cities, these lizards perch on fences, walls, ornamental trees, and the sides of buildings, often at eye level. If you stand quietly near a lamppost on a warm evening in San Juan, you may see a crested anole flatten itself against the pole and run upward, as if gravity were a mild suggestion. That performance looks effortless because the animal is built to make it look that way.

Anoles climb using an adhesion system that resembles a high-tech material more than a typical foot. Beneath the toes are expanded pads, and across those pads run rows of tiny scales called *lamellae*. Each lamella bears even smaller hair-like structures (setae), which branch into microscopic tips. When pressed against a surface, those tips get close enough for weak intermolecular attractions to add up. The force is not glue and not suction; it is the collective effect of countless intimate contacts. The same kind of physics makes a gecko's foot famous, but anoles are not geckos. Their pads are smaller, their ecology is different, and their evolutionary story has its own twist.

City surfaces change what matters in that adhesion system. Rough bark offers mechanical interlocking: claws and skin can hook into pits and ridges. Smooth paint denies that. On a smooth wall, the quality of attachment depends much more on pad area, lamellae number, and how well the toe can be oriented to maximize contact. Even posture changes: the angle at which a lizard approaches a perch alters which parts of the foot bear load and how the pad peels away when it moves. In a forest, these details can be forgiving. On a glassy pillar, they become decisive.

Kristin Winchell and colleagues (2016) studied crested anoles living in Puerto Rican cities and nearby forests and linked their habitat use directly to their anatomy. Urban lizards were found more often on broad, smooth, artificial substrates, while forest lizards used narrower, rougher, more varied perches. The two groups also differed morphologically in ways that make biomechanical sense: urban individuals tended to have relatively longer limbs and more subdigital lamellae. Longer limbs can increase stride length and help with speed on broad surfaces, where stability is less constrained by narrowness. More lamellae generally means more potential contact points, which can translate into better adhesion on smooth substrates.

The beauty of this system is that it invites an almost mechanical way of thinking about natural selection. Imagine a city block as a testing apparatus. Fence posts, stucco walls, and metal railings are the standardized surfaces. A lizard that clings well can use these structures as highways, escaping predators, finding mates, and expanding its territory. A lizard that clings poorly is forced downward into more exposed ground habitats or into the few remaining rough-barked trees. Survival and reproduction, in such a setting, become entangled with surface engineering.

A lizard's foot is not a single trait, and neither is climbing. The number of lamellae matters, but so do their size, shape, and spacing. Toe pad width matters, but so does the stiffness of the toes, which influences how evenly the pad presses against a surface. Limb length matters, but so does limb muscle arrangement and joint flexibility. Even the skin on the underside of the foot can affect how it conforms to microtexture. Yet when natural selection is strong and consistent, it does not need to perfect every detail. A modest increase in lamellae number, repeated across generations, can yield a meaningful shift in performance.

A diagram can make the idea concrete: not the microscopic details that require an electron microscope, but the macroscopic layout of the toe pads and lamellae rows that a careful observer could see with a hand lens.

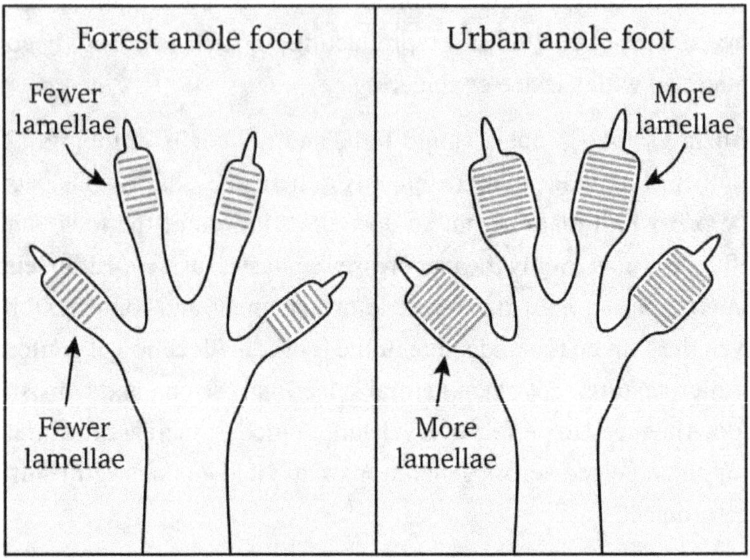

Figure 1. Split-view schematic of toe-pad differences between forest and urban *Anolis cristatellus*: urban lizards tend to have broader pads with more lamellae, improving contact on smooth, vertical substrates.

If the city's glass and paint are the challenge, why does the response look so specific? Because the selective pressure is specific. A lizard does not need to become "better at climbing" in some abstract sense. It needs to avoid falling on the surfaces it actually uses. When selection is tied to a repeated performance demand, the anatomy that supports that performance becomes a predictable target.

There is a temptation to treat such changes as mere flexibility within a lifetime. After all, muscles grow with use; bones remodel under load; developmental conditions can alter body proportions. This is *phenotypic plasticity*: the ability of a single genotype to produce different phenotypes depending on environment. Plasticity is real, and it often matters in cities, where conditions can change quickly. Yet the key question is whether urban anoles are simply shaped by living on walls, or whether populations have evolved genetically.

One way to pry those apart is to raise individuals from different habitats under similar conditions and see whether differences persist. Winchell's work combined field measurements with evidence consistent with at least some heritable component to the morphological divergence. The point is not that every aspect is genetic or that plasticity is irrelevant. The point is that the city can create conditions under which genetic differences accumulate fast enough to be detected over surprisingly short times.

Time is the shocking element. Urbanization in Puerto Rico, as in many places, is recent on an evolutionary scale. A human lifetime can witness the construction of new neighborhoods, the paving of roads, the replacement of gardens with parking lots. For a lizard with a generation time measured in a year or so, that is many generations. In the same way that heat islands can shift thermal tolerances in a few decades for small organisms, the built environment can shift the biomechanics of movement over similarly brief spans.

The physical argument is straightforward, but evolution is rarely a single-factor story. City life also changes predators, prey, parasites, light cycles, noise, and the arrangement of vegetation. How can one be confident that the differences in limbs and toe pads relate to surfaces rather than something else?

One methodological answer is replication. If one compares a single urban population to a single forest population, any difference could be an accident of history. The more powerful design repeats the comparison across multiple urban—forest pairs and asks whether the same traits and the same parts of the genome show similar divergence each time. Genome-wide studies of urban anoles have used such replicated designs across different regions of Puerto Rico, helping to separate the "urban effect" from the background pattern of geographic relatedness. When independent urban populations show parallel genetic signals associated with urban living, the case for adaptation strengthens. It becomes less plausible that all the differences are due to chance colonization events or peculiar local histories. Natural selection leaves a certain kind of fingerprint: not a single dramatic mutation, but a repeated tilt in allele frequencies in the same direction across repeated environments.

There is a satisfying concreteness to the anole story because it puts evolution at street level. The evolutionary biologist's measurements—limb segment lengths, lamellae counts, sprint speed, clinging force—can be traced back to things anyone can recognize: a painted wall, a metal fence, a windowpane. That concreteness helps resolve a common misunderstanding. Evolution is often pictured as a response to grand, slow forces: ice ages, continental drift, the rise of mountains. Cities show how it also responds to the geometry of the ordinary.

A historical echo makes the point. When Charles Darwin wrote about natural selection, he repeatedly reached for analogies in domestication: pigeon fanciers selecting particular traits, farmers choosing breeding stock. He did that because selection is easiest to understand when it is visible and when the criteria are clear. Urban architecture, unintentionally, plays a similar role. It creates a standard that animals must meet. No one decides that a lizard should have more lamellae. Yet the city, by building smooth vertical surfaces everywhere, sets the exam paper.

The exam is not only about climbing. Consider a lizard moving across a broad wall in full sun. It must maintain adhesion, but it must also regulate body temperature, avoid predators, and communicate with rivals. A longer limb can improve speed on broad perches, but longer limbs can also change how a lizard displays, how it turns, and how it escapes on narrow branches. An expanded toe pad might improve grip, but toe pads can carry costs too: they add mass at the distal end of the limb, potentially affecting swing dynamics; they may be more prone to wear; they might alter sensitivity. Evolution is not "designing" the perfect city lizard; it is balancing a portfolio of trade-offs under a new set of constraints.

One of the most engaging parts of the Puerto Rican system is that the lizards themselves conduct demonstrations of those trade-offs in real time. Watch an anole on a tree and it often moves with a certain confidence, using claws and friction. Watch one on a glass window and the movement becomes more deliberate: toes splayed, pads pressed, the body close to the surface. The lizard is solving a mechanical problem with its body. When the solution works, it looks like ease. When it does not, the slip is immediate and honest.

If the city has been a furnace for thermal physiology, it is also a sculptor of contact mechanics. The sculpting is not abstract. It is written in how many lamellae a toe carries, in how wide the pad spreads, in how long a limb reaches. It is written, too, in the everyday routes that animals take. The city is full of corridors, but they are corridors made of materials that reward certain bodies.

There is a small human poignancy in this, because the lizard's success is tied to our preferences. We like clean lines, smooth paint, polished stone, glass facades. We build the kind of surfaces that are easiest to wash and most resistant to weathering, and then we are surprised when the local fauna sorts into winners and losers. A rough-barked tree is a complex habitat; a stucco wall is a simplified one. The simplification does not remove nature; it changes the terms on which nature can persist.

On a warm evening, a crested anole perched on a concrete wall is not merely "coping with the city." It is using the city as a habitat in the most literal way: making a living on the geometry we have erected. Its toe pads spread, lamellae contacting the surface in neat rows, turning a human-made plane into something climbable. The wall does not care whether the lizard lives or dies, but the lizard's descendants will carry, in millimeters and in scale counts, the evidence that it mattered.

4.3 Evolution Underground

In the middle of the twentieth century, Londoners learned to treat the Underground as more than a transport system. During the Blitz, stations became shelters. People slept on platforms beneath posters and bare bulbs, waiting for the all-clear, trying to make an ordinary night inside a concrete burrow. Those nights also supplied a different kind of intimacy with city wildlife: a mosquito that did not behave like the familiar outdoor sort. It bit readily, even in winter. It seemed perfectly at home underground, as if the tunnels themselves had bred it.

The story that grew around this insect became one of the most repeated modern fables of evolution: a new form, perhaps even a new species, born in a century of steel and soot. There is a reason it was so irresistible. It offered a clean moral about human construction producing immediate biological consequences. It made speciation feel like something you could ride to work.

The mosquito is real. What has changed is the interpretation of what it represents.

The starting point is a species complex, a cluster of closely related forms that look deceptively alike but differ in habits that matter to survival and reproduction. The *Culex pipiens* complex includes mosquitoes that are common across temperate regions, often breeding in standing water and feeding largely on birds. Within that complex sits a form traditionally called *Culex pipiens f. molestus*, associated with enclosed human-made habitats: basements, sewers, tunnels, and the warm cavities that cities inadvertently provide.

What made the "Underground mosquito" famous were not minor genetic differences on a laboratory readout but a suite of biological shifts that sound, even to a non-specialist, like the beginnings of a new way of life. The molestus form tends to bite mammals more readily, including humans, rather than focusing on birds. It can reproduce in confined spaces where swarming flight may be constrained. It often remains active year-round rather than entering a winter dormancy called *diapause*. Each of these traits is a kind of key, opening doors that are locked to a more seasonal, bird-oriented mosquito.

Cities do not merely add hazards like heat and smooth walls; they also add *barriers*. A road can be a boundary. A rail line can be a moat. A patch of dry pavement can be a desert to a creature that depends on damp shade and cover. In the case of subterranean insects, infrastructure can do something even stronger: it can take a population and tuck it into a physically separated world, with limited contact to its relatives aboveground.

That separation matters because speciation is, at heart, the evolution of reproductive separation. A biologist might define speciation as the process by which one lineage becomes two lineages that do not freely exchange genes. The phrase *reproductive isolation* captures the practical side: if individuals from two groups rarely mate, or mate but produce low-fitness offspring, then the groups begin to travel separate evolutionary paths.

Isolation does not have to be dramatic. It can begin as a slight reduction in gene flow, the trickle of interbreeding that is cut to a drip.

Over time, selection and chance can widen the gap. Selection can push traits in different directions because the environments differ. Chance can push populations apart through *genetic drift*, the random fluctuation of gene frequencies that becomes especially influential when populations are small.

The London Underground offered, at least superficially, a textbook "allopatric" scenario: a population cut off by geography (in this case, by tunnels, stations, and enclosed breeding sites) and forced to adapt to a distinct environment. In 1999, Kathryn Byrne and Richard Nichols published a study in *Heredity* that gave the story a firm empirical footing. They compared mosquitoes collected from Underground sites with those from nearby surface locations, sometimes only a short distance away. The genetic patterns were striking: subterranean and surface mosquitoes in London were distinct, with little evidence of gene flow between them at the sampled sites. Underground populations showed reduced heterozygosity, a signal consistent with colonization by a small number of founders and subsequent isolation. Breeding experiments also suggested reduced compatibility between the underground form and surface populations.

For a time, this looked like a rare gift: a speciation process unfolding in an environment whose construction date could be read off engineering history. The Underground began in the 1860s, expanded rapidly, and by the early twentieth century had become a sprawling subterranean habitat. If the molestus form had truly originated within those tunnels in under two centuries, it would have been a spectacular example of fast evolutionary divergence.

The deeper reason it seemed plausible is that cities can impose very strong selection. A mosquito that cannot reproduce without a cold winter pause might fail underground, where temperature is buffered and winter never fully arrives. A mosquito that depends on outdoor mating swarms might struggle in confined airspace. A mosquito that prefers birds might not thrive if birds are scarce below street level but humans are abundant. The underground habitat appears to reward a particular package of traits, the kind that makes a mosquito an unwelcome companion in basements and shelters.

Yet the very neatness of the story carried a danger. If a result is famous, it tends to be simplified in retelling. "The Underground mosquito evolved in London" can slide into "the Underground mosquito evolved because the Underground was built," and then into "human infrastructure created a new species." Each step sounds only slightly bolder than the last, but the meaning shifts.

A major reassessment arrived with a 2025 *Science* paper, which analyzed population genomic data from a large set of contemporary and historical samples. The conclusion was blunt: the popular hypothesis that molestus originated belowground in London less than 200 years ago does not fit the genomic evidence. Instead, the study argued that molestus first adapted to human environments aboveground in the Mediterranean or Middle East over more than a millennium, plausibly in association with ancient agricultural societies. London's tunnels, on this view, did not *create* molestus. They *collected* it.

This is not a demotion of evolutionary interest. It is, in some ways, a richer story, because it forces attention onto a neglected idea: *preadaptation* (or, more carefully, exaptation). Sometimes traits that seem tailor-made for a new environment were shaped elsewhere, under earlier conditions, and then happen to work extremely well in a novel setting. A mosquito lineage that evolved to exploit enclosed, human-associated habitats in the past would find the modern city's basements, sewers, and subways to be a ready-made archipelago of suitable sites.

The revision also changes what "fast" means. The divergence of molestus from other forms may not be as recent as the Underground itself, but the *urban interface* it represents could still be ancient and continuous. Humans began storing water, irrigating fields, building granaries, and shaping wet microhabitats long before the first subway. In other words, the evolutionary pressure of "human environments" is older than modern cities. The city merely concentrates and standardizes it.

So what remains of the Underground mosquito as an example of speciation through infrastructure-based isolation?

Plenty, once the emphasis shifts. The most instructive feature of the Underground is not that it conjured a new insect overnight, but that it creates sharp boundaries in gene flow at human scales. Mosquitoes are mobile, yet populations can be structured by the availability of breeding sites and by the geometry of the built environment. A tunnel system is not just a corridor; it is also a filter, connecting some sites while sealing off others. Even a species that can fly may rarely cross from a warm subterranean breeding pool to an aboveground population if the intervening landscape is dry, hostile, or simply irrelevant to its life cycle.

The Underground also illustrates how reproductive isolation can be partial, complicated, and geographically dependent. Byrne and Nichols noted evidence that in southern parts of the species' range, traits of molestus and pipiens can mix more readily. Reproductive barriers are not always absolute walls; they can be leaky membranes. That matters because "species" is not always a yes-or-no category in the early stages of divergence. A form can be behaviorally distinct, ecologically specialized, and genetically differentiated while still capable of hybridization under some conditions.

That ambiguity is not a flaw in evolution; it is one of its signatures. Nature rarely presents crisp categories in the middle of a process.

A second lesson comes from how easily we confuse selection with separation. Urban heat islands can drive local adaptation because populations experience different temperatures and survive differently. Underground habitats can drive divergence for the same reason, but there is an additional force at work: isolation itself. When a small number of founders colonize a new habitat, the genetic composition of that founding group can differ by chance from the source population. Over time, that difference can grow through drift, even without strong selection. In a confined subterranean habitat, population size may also fluctuate, amplifying drift further.

The challenge for researchers is to disentangle these forces: which differences reflect natural selection in a new environment, and which reflect demographic history? This is not a technical quibble. If one mistakes drift for adaptation, one can end up inventing just-so stories about why a trait "must" be useful underground, when in fact it may simply have hitchhiked along with a founder event.

Work on urban mammals makes the point vivid. In New York City, white-footed mice live in parks that function like habitat islands, separated by roads and buildings. Jason Munshi-South and colleagues (2010) showed that populations in different parks can be genetically distinct even across only a few kilometres, implying that the urban matrix is a meaningful barrier to dispersal. Later genomic studies emphasized the importance of modelling demography when searching for signals of selection in such fragmented populations, precisely because bottlenecks and isolation can mimic the genetic fingerprints of adaptation.

The Underground mosquito sits at the intersection of these ideas. Its story has always involved both ecology and isolation: a warm, enclosed habitat with a different menu of hosts, and a set of barriers that reduce interbreeding with aboveground mosquitoes. The 2025 genomic reassessment suggests the lineage's core adaptations to human-associated enclosures may have deeper roots, but the Underground still acts as a sorting device, maintaining those adaptations in a modern setting and shaping how they interact with surface populations.

For a mosquito, the difference between biting birds and biting humans is not a preference; it is a route through an ecosystem. Bird-feeding mosquitoes mostly circulate pathogens among birds. Mammal-feeding mosquitoes can connect bird infections to humans if they bite both, or if hybrid populations bridge the behavioural divide. That public-health dimension has helped keep the Underground mosquito in the spotlight, and it underscores a broader point about fast evolution in cities: the traits that change are often the ones that place humans directly inside the evolutionary story, not as observers but as hosts, builders, and sources of selection.

The romance of a brand-new species evolving between the opening of a railway line and the lifespan of a newspaper archive is harder to defend now. But the Underground still offers something rare: a clear view of how human infrastructure creates biological islands and corridors. The city is not one environment but many, stacked and stitched together. Aboveground, the heat and the smooth surfaces impose their own filters. Belowground, darkness, humidity, stable temperatures, and enclosure reshape life in a different register.

In a sense, the mosquito's fame came from seeming to prove that evolution can keep pace with construction. The deeper truth is that evolution has been keeping pace with human environments for a very long time, waiting, so to speak, in the wings of our architecture. When a tunnel opens, it does not merely create a space; it connects to an older ecological theme: water held in darkness, warmth without winter, hosts gathered in dense numbers. A mosquito that can exploit that theme will appear as if it were made for the Underground, because in important ways it was made for us.

Chapter 5

Chemical Warfare

Pollution is typically viewed as a force of destruction that leaves only ecological wastelands, yet for many species, it acts as a potent sculptor of life. This chapter reveals how the byproducts of human industry have inadvertently launched a global experiment in natural selection, forcing organisms to rewrite their genetic code to survive chemical gauntlets. We will discover that evolution in the Anthropocene is not just about survival of the fittest, but often involves striking dangerous bargains with biology itself.

5.1 The Moth and the Smokestack

On a soot-stained wall, a moth is either a smudge or a signal flare.

That simple contrast made the peppered moth, *Biston betularia*, one of the best-known creatures in evolutionary biology. It is an unassuming insect—about the size of a thumb, night-flying, and usually easy to ignore. Yet for more than a century it has carried an outsized burden: to show that natural selection is not merely an ancient force inferred from fossils, but something that can reshape a living population quickly enough to track a changing human environment.

The story begins with a palette change. In much of Britain, the Industrial Revolution did not only transform work and wealth; it recolored the countryside. Coal smoke and sulfur-laden fumes blackened buildings and tree trunks and killed many of the pale lichens that had once speckled bark with grey-green crusts. If you stood under a streetlamp in a manufacturing town in the late nineteenth century, the world could look as though someone had dragged a charcoal stick across it.

The peppered moth came in two main paint jobs. The "typical" form was pale, peppered with black spots—an effective disguise on lichen-covered bark. The melanic form, called *carbonaria*, was much darker, nearly black. Melanism simply means an increase in dark pigment (melanin) in the skin, scales, feathers, fur, or wings. It can arise for many reasons across the animal kingdom. What made this case electric was the timing: records from collectors suggest that the dark form was rare before industrial pollution, then surged in polluted regions, then declined again as air quality improved.

A good story, though, is not automatically a good explanation. The temptation is to treat the peppered moth as a morality tale with an obvious moral: soot makes trees dark; dark moths hide better; birds eat the moths they can see; therefore dark moths become common. It is tidy enough to fit on a classroom poster.

Nature is rarely so polite. A population can change for reasons that have nothing to do with camouflage. Individuals can move in from elsewhere. Dark forms might be hardier for other physiological reasons. Perhaps the insects rest in places we do not expect. Perhaps the experiments were too artificial. If the peppered moth is going to stand for evolution on human timescales, it has to withstand the kind of scrutiny we would apply to any claim about cause and effect.

The heart of the evidence rests on a simple logic. Natural selection requires three ingredients: variation (individuals differ), inheritance (offspring resemble parents), and a link between those differences and survival or reproduction. The peppered moth supplies the first two easily: the light and dark forms are distinct, and the trait is heritable. The real question is the third ingredient: did the environment—made darker by pollution—actually change which moths lived long enough to reproduce?

A celebrated set of mid-twentieth-century field experiments, associated with Bernard Kettlewell, tried to answer that question directly by releasing moths of different colors in contrasting environments and measuring which ones were eaten more often. The basic idea was reasonable and remains a standard approach in ecology: if predators remove one form more than another, and if that predation matches the camouflage context, then selection is operating in the predicted direction. But the details became controversial over the decades. Critics questioned whether moths naturally rest on exposed tree trunks, whether the released moths behaved normally, whether the experimental sites were representative, and whether the presentation of results in popular accounts oversimplified what was actually done.

Science can absorb these arguments without collapsing. A classic example can be both broadly correct and messier than the simplified version. In the peppered moth case, later work strengthened the predation explanation by using more natural observations and by repeating experiments with improved designs. Michael Majerus, who devoted much of his career to this system, carried out careful studies in the early 2000s that recorded differential bird predation in real woodland settings. The conclusion was not that every earlier criticism was irrelevant, but that the main mechanism—predators preferentially removing the more conspicuous morph—was real and strong enough to drive the observed frequency changes.

That still leaves a deliciously modern twist. For many people, the peppered moth feels like an old case: Victorian smoke, Edwardian collectors, mid-century experiments. The surprise is that it now has a molecular fingerprint. In 2016, Arjen Hof and colleagues identified the specific genetic change responsible for the *carbonaria* form: a large transposable element insertion in a gene called *cortex*. A transposable element is often described, informally, as a "jumping gene"—a stretch of DNA that can copy or move itself to new locations in the genome. Most of the time such insertions are harmful or neutral. Occasionally, one lands in a place where it tweaks how a nearby gene is regulated, and the organism looks or behaves differently.

In Lepidoptera (the insect group that includes moths and butterflies), *cortex* is involved in wing pattern development. The insertion sits in an intron (a non-coding region within a gene that is removed from the final messenger RNA) and appears to change when, where, or how much *cortex* is expressed during the early development of the wing. Instead of a gradual accumulation of tiny pigment changes, the melanic form is linked to a single identifiable genetic event with a large visible effect. That does not mean evolution is always driven by big mutations—often it is not—but it is a reminder that rapid evolution can sometimes hinge on a surprisingly discrete piece of genetic luck.

This discovery also allowed geneticists to ask a new question: did the dark form arise many times, in many places, as pollution spread, or did it arise once and then spread? By looking at the surrounding DNA (the haplotype) associated with the melanic allele, the evidence points to a single origin in Britain, with an estimated origin date around the early nineteenth century. One mutation, arriving at the right historical moment, became a ticket that natural selection could punch again and again.

At this point it is worth slowing down to say what is really meant by "quickly." In everyday speech, evolution sounds geological: continents shifting, dinosaurs fading, deep time. But the relevant clock for natural selection is not years; it is generations. A moth can have one or more generations per year, depending on climate and ecology. If a trait changes in frequency over a few decades, that can represent dozens of generations—enough for selection to do substantial work, especially if the selective pressure is strong.

Industrial melanism provided something even rarer than a rapid rise: a rapid fall. Beginning in the middle of the twentieth century, coal use began to change and air-quality regulations took hold. In Britain, the Clean Air Act of 1956 (spurred by the lethal Great Smog of London in 1952) and later measures reduced smoke and sulfur dioxide emissions. Lichens began to return to tree trunks in many regions. Buildings and bark lightened. And the frequency of the dark moth declined, in some places dramatically, over a few decades. Evolution did not simply sprint in one direction and stop; it reversed course when the selective landscape reversed.

That bidirectional pattern is unusually persuasive because it is hard to fake with a single confounding factor. Migration could, in principle, bring in light moths from cleaner areas. But the decline happened across multiple locations and tracked environmental improvement in a way that makes pure migration an incomplete story. The most compelling interpretation is also the simplest: when the background lightened, the camouflage advantage shifted back to the pale form, and predators once again did their quiet accounting at dusk.

A time-series view makes the point more clearly than any rhetorical flourish. The chart below is a schematic overlay of two curves: an atmospheric measure (sulfur dioxide, a major component of industrial air pollution) and a biological measure (the frequency of the melanic *carbonaria* morph). The exact numbers vary by site, sampling method, and region; the value lies in the shape. Pollution falls first. The moth population follows, but not instantly. Even when selection is strong, populations can lag behind environmental change because genes must be passed through generations, and because landscapes are patchy: a cleaner year does not repaint every tree overnight.

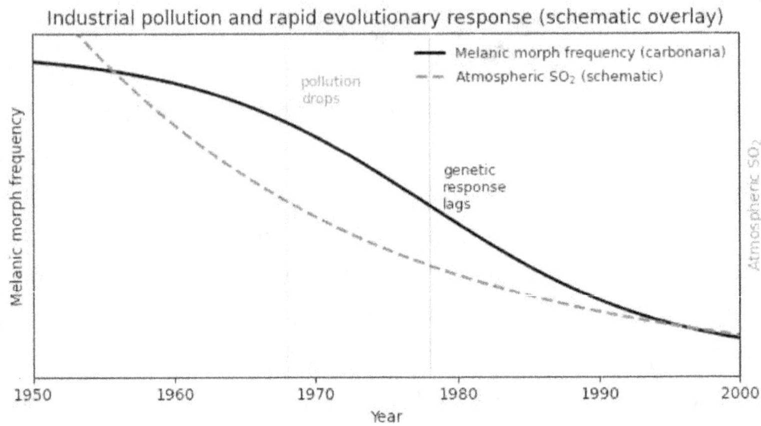

The lag is not a flaw in the story; it is part of what makes it realistic. Evolution is not clairvoyant. A population does not reconfigure itself in anticipation of new regulations or cleaner fuels. It responds, generation by generation, to what kills and what spares.

One might still wonder why camouflage should matter so much. Would a small mismatch between moth and bark really decide life or death? A useful way to imagine the selection pressure is to picture a bird feeding its chicks. Many small songbirds hunt visually. A moth resting by day is a packet of calories. If a bird can spot one morph twice as easily as another against a particular background, the more visible morph may suffer disproportionately high mortality. Over the course of a season, those differences can translate into a substantial shift in which individuals survive to mate.

This kind of selection is especially potent because it is so ordinary. There is no need for a rare catastrophe, no special predator, no elaborate ecological chain reaction. The selecting agent is the daily decision-making of common animals: glance, peck, swallow. That ordinariness is also why the peppered moth still feels relevant in a world where smokestacks are cleaner and many pollutants are regulated. Human environments routinely alter the backgrounds against which animals must hide, hunt, or signal. A city park, a roadside verge, a forest fragmented by development—each creates new visual and chemical contexts. Predators and prey do not need to "know" what changed; their success rates change automatically.

The molecular detail adds another layer of intimacy. When people say evolution is "just a theory," they sometimes mean it feels abstract, like an opinion about the past. But here is a literal insertion of DNA in a specific gene, in a specific place, producing a specific wing color. That piece of DNA rose and fell in frequency as the human atmosphere darkened and then cleared. It is difficult to ask for a more concrete connection between industrial history and the arithmetic of inheritance.

It also helps to be honest about what remains uncertain. The strongest evidence supports bird predation on differently camouflaged moths as the dominant driver. Yet dominance is not exclusivity. Some researchers have raised the possibility that pollution might have had other effects: on the moths' physiology, on their resting behavior, on their susceptibility to disease, on the microclimate of their habitats. The post-1970s decline in melanism has also prompted questions about the roles of migration and gene flow, especially in landscapes where clean and polluted patches sit close together. In other words, even in an icon, the science does not end with a single arrow labeled "soot $\rightarrow$ dark moths." It becomes a more textured picture: selection acting strongly, but within a living, moving population.

There is a human anecdote often told about the peppered moth that captures the emotional core of the case. In the smoky decades when *carbonaria* was common, collectors could supposedly take a rest in an industrial woodland, glance at a tree trunk, and pick off dark moths like bits of bark. In cleaner times, the same gesture would find mostly pale moths. Whether or not the scene is as effortless as the anecdote suggests, it points to something true: the shift was not a subtle statistical change visible only in a lab. It was a change you could hold between your fingers.

The peppered moth does not prove that every trait in nature is perfectly adapted, or that evolution always proceeds quickly, or that human impacts will always produce neat, reversible patterns. What it demonstrates is more precise and, in a way, more unsettling: when humans change the world, we do not merely move organisms around or reduce their numbers. We change which genes succeed. A smokestack does not only stain stone; it rewrites the odds of survival, and in doing so it can reach into the genome of an ordinary moth and tip a population toward black or toward white.

In a clean woodland, a pale peppered moth resting among returning lichens looks almost painted into the bark, as if it belongs there in the most literal visual sense. A dark moth in the same place is a crisp silhouette, the kind of shape a bird's eye was built to notice. In that contrast lies a quiet lesson about speed: evolution can be fast enough to keep pace with policy, fast enough to register the difference between a dirty decade and a cleaner one, fast enough that the atmosphere above a city can leave a signature in the genes of an insect clinging to a tree.

5.2 Convergent Concrete Jungles

A city looks chaotic from the inside: traffic lights, delivery vans, cracked sidewalks, park lawns shaved to the same height, and trees that grow in square holes cut out of asphalt. To an organism trying to make a living, cities are not chaotic at all. They are patterned, repetitive, and strangely similar across continents. Pavement and brick soak up heat in the day and leak it back at night. Salt is scattered on roads in winter. Soil is compacted, disturbed, and imported. Water arrives in pulses from sprinklers and drains away through gutters. Herbivores that would once have kept plants on their toes often vanish, replaced by a narrower cast of urban survivors.

From an evolutionary perspective, a city is an inadvertent experiment built at enormous scale, then repeated hundreds of times with local variations. Each repeat poses a question that is both scientific and philosophical: if similar problems appear in different places, do organisms solve them in similar ways?

That question has a name. *Parallel evolution* describes a pattern where populations facing similar selection pressures—the forces that make some individuals leave more descendants than others—evolve similar traits independently. It is not magic, and it is not destiny. It is an empirical claim about repeatability, the extent to which natural selection can be predictable when the environment is predictable. Cities are among the few human-made environments that are both intense and replicated enough to let the question be asked with real statistical muscle.

One of the best natural probes of this idea is humble enough to be stepped on.

White clover, *Trifolium repens*, is the familiar three-leaved plant that threads through lawns and parks. It is not native everywhere it now grows, and it thrives in the disturbed habitats humans create: mowed grass, soccer fields, roadside verges, gaps between paving stones. Clover also carries a small chemical weapon. Some clover plants, when their tissues are crushed, produce hydrogen cyanide. Cyanide is the same molecule that appears in spy novels and warning labels, but in plants it functions as a deterrent: a small animal that chews clover can get a sharp, unpleasant dose.

This defense is called *cyanogenesis*—literally "cyanide creation." It is a two-part system. One set of compounds stores the potential for cyanide, another set of enzymes releases it when the plant is damaged. The system is effective, but it is not free. Maintaining the ingredients and the machinery costs resources. In cold conditions, the mechanism can become risky because damaged tissues may release cyanide when the plant is already stressed, and freezing can alter cell structure in ways that make accidental release more likely. The details of the physiology can be intricate, but the ecological consequence is straightforward: cyanogenesis is useful in some environments and disadvantageous in others.

That makes it an ideal trait for studying rapid, local adaptation. Adaptation means a trait becomes more common because it improves survival or reproduction in a given environment. The core signature is not merely that a trait varies, but that it varies *systematically* with the conditions that should make it beneficial or costly.

Now place clover into the urban mosaic. Cities create a distinctive microclimate: the *urban heat island* effect. Concrete, asphalt, and buildings absorb and retain heat, raising temperatures relative to nearby rural areas, especially at night. Snow can melt sooner. Frost can be less severe. At the same time, cities often reduce the abundance of large grazing herbivores, and even where small herbivores persist, the community of chewers and suckers may differ. In a neatly mowed park, the most common form of "herbivory" may be a lawnmower.

Given those ingredients, cyanogenesis becomes a testable story. If cyanide is mainly a defense against herbivores, then plants in places with fewer herbivores might be expected to lose it over time. If cyanogenesis is costly under cold stress, then warmer urban environments might maintain it. The point is not to pick one simplistic prediction and declare victory; it is to see whether nature sorts the trait in a way that matches the balance of costs and benefits in each city.

A global view arrived when James Santangelo, Rob Ness, and collaborators assembled an astonishing dataset (2022): more than one hundred thousand clover plants sampled across thousands of populations in 160 cities worldwide. The scale matters because cities differ. Some are warm and wet, others hot and dry. Some have lush parks, others are largely hardscape. If parallel evolution is real, it should appear as a *cline* repeated across many cities.

A cline is a gradient: a systematic change in a trait across space. It can occur along a coastline, up a mountainside, or—in this case—from a city center out into the surrounding countryside. If you stand in the dense core of a city and move outward, you often experience a change in temperature, vegetation cover, soil moisture, and land use. A cline in clover cyanogenesis would mean that the probability of a plant producing cyanide changes in a consistent way along that urban—rural transect.

The striking result was that such clines are common but not universal. Roughly half of the sampled cities showed clear urban—rural clines in cyanogenesis. In many places, urban clover populations had a lower frequency of cyanide-producing plants than rural populations. That is exactly the kind of repeated pattern that suggests natural selection is acting in similar ways across disparate locations: Tokyo is not Toronto in language, architecture, or history, but it can present a clover plant with surprisingly similar problems.

This is where the "convergent concrete jungle" idea earns its name. Convergent evolution refers to different lineages evolving similar traits. Parallel evolution is a close cousin, usually describing similar evolution in closely related populations or species. In either case, the fascination lies in the same place: independent histories arriving at similar endpoints.

It would be comforting, and wrong, to say that cities simply erase cyanogenesis everywhere. Variation among cities is not a nuisance; it is information. Some cities have strong drought stress in urban cores, with dry soils and sparse vegetation. In such settings, the cost of cyanogenesis may be different than in a city with irrigated lawns and shaded parks. Santangelo and colleagues found that environmental differences among cities—including vegetation cover and indicators of drought stress—help explain why some cities show strong clines and others do not. Parallel evolution, in other words, is not a claim that outcomes are identical everywhere. It is a claim that when selective pressures align, evolution often aligns too.

A useful way to imagine this is to think of cities as sharing a set of recurring features, like a template, but with local edits. The template includes warmer nights, disturbed soils, fragmented green spaces, and altered communities of herbivores and microbes. The edits include climate, irrigation, park management, and the surrounding regional landscape. Evolution reads both: it responds to the common pattern and the local footnotes.

There is a second layer of interest in clover because the trait is not a vague, hand-wavy "hardiness." Cyanogenesis is genetically tractable. It depends on the presence or absence of functional components in the biochemical pathway. That structure means the trait can change relatively rapidly. If plants lacking cyanogenesis have a consistent advantage in certain urban environments, the genetic variants that disable cyanogenesis can spread over surprisingly few decades.

This is where the earlier story of industrial melanism casts a long shadow without needing to be retold. In the moth, a visible trait tracked soot and then clean air. In clover, a chemical trait tracks the urban gradient. In both cases, the pace is set by generations, not by the human calendar, and the selection pressure is not exotic. It is the summed consequence of many small encounters: birds searching bark; insects nibbling leaves; frost biting tissue; drought constraining growth.

Cities add something new: replication. A smokestack is local. A city is global.

Historical anecdotes can make this feel less abstract. The idea of cities as distinctive climates is not new. Nineteenth-century observers in London remarked that fog and smoke made the city feel like a world apart. Charles Dickens's descriptions of urban air were not meteorological measurements, but they captured a truth that modern sensors quantify: city air and city temperature are often measurably different from those of the countryside. Urban heat islands, now mapped by satellites and weather stations, formalize what residents have long felt on summer nights when the countryside cools and the city does not. For an organism with no air conditioning, those differences are not just comfort; they are selection.

The most intriguing aspect of urban evolution is how often it draws lines across a space that humans treat as seamless. People drive from downtown to suburb to farmland without thinking of it as a biological gradient. A plant experiences it as a sequence of habitats stitched together: hotter pavement edges, watered lawns, shaded park understories, roadside salt spray, compacted soils, and remnant meadows beyond the city.

That stitched landscape makes clines almost inevitable if selection differs between the patches and if gene flow—the movement of genes through pollen and seed—is not so overwhelming that it homogenizes everything. Gene flow can blur adaptation by constantly reintroducing variants from other environments. The existence of urban—rural clines implies that selection is strong enough to maintain differences despite mixing, or that the mixing itself is limited by the patchy geography of urban green spaces.

Clines also prompt a subtle but crucial question: are we seeing genetic evolution or merely flexibility? Many organisms show *phenotypic plasticity*, the ability of a single genotype to produce different traits in different environments. A plant may grow thicker leaves in sun than in shade without any genetic change. Plasticity can be adaptive, and it can arise quickly, within a lifetime.

Distinguishing plasticity from evolution requires additional evidence: common-garden experiments, genomic data, or careful inference. In the clover case, the trait has a clear genetic basis and is not simply a leaf-shape response to light. The global pattern, repeated across many cities, also points toward evolution rather than purely plastic adjustment. Yet plasticity is never off the table in urban environments, and it interacts with evolution in ways that can either speed adaptation (by helping organisms survive long enough for selection to work) or slow it (by masking genetic differences).

Even without plunging into technical details, the key insight is accessible: cities do not merely test organisms; they may also *train* them. The same urban features that impose selection—heat, disturbance, altered predators—can also induce developmental responses. The result is a living mixture of immediate flexibility and slower genetic change, both operating on a timescale humans can observe.

Plants are particularly interesting here because their relationship with place is intimate. A moth can fly away. A clover plant cannot. It must endure the soil it germinates in, the mower that passes over it, and the dog that runs across it. That immobility makes plants sensitive recorders of local conditions, and it helps explain why urban clines can be so crisp.

The fascination of parallel evolution is not merely that it makes a nice pattern on a map. It forces a deeper question about constraint. If clover repeatedly loses cyanogenesis in city centers, is that because losing it is the best solution, or because it is the easiest solution available given clover's genetic toolkit? Evolution is not an engineer with unlimited options; it is a tinkerer working with existing parts. In clover, disabling a chemical defense pathway may be a relatively simple genetic step with a clear energetic payoff. In another species, the analogous step might be complicated, or might break something else.

That is why urban evolution is sometimes predictable and sometimes not. A repeated selection pressure does not guarantee a repeated response. It increases the odds that we will see similar shifts, but the outcome still depends on the genetic variation available, the costs of different solutions, and the local ecological context. Even within clover, only about half the cities show the neat clines that make parallel evolution look like a law of nature. The other half are reminders that ecology has many dials, and cities do not all turn them the same way.

A concrete way to hold this in mind is to imagine three city centers—any three: the core of a dense North American metropolis, a Japanese city with compact green spaces, a European city with older parks and different mowing regimes. In each case, clover populations can be sampled in rings radiating outward. If cyanogenesis declines toward the center in all three cities, that is a parallel cline. The diagram below depicts this schematically, using color to represent the frequency of cyanide-producing plants: red for high cyanogenesis, blue for low. The key feature is not the geography but the repeated radial gradient.

Parallel urban–rural clines in white clover (schematic)

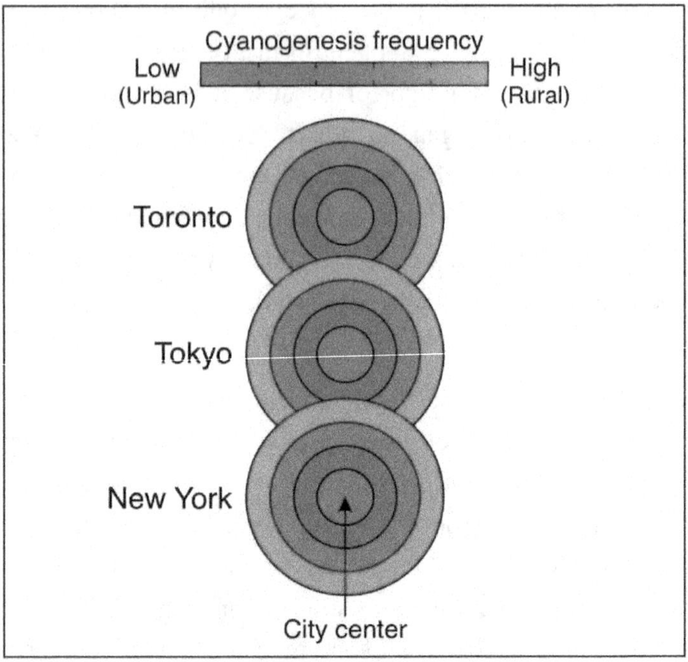

Blue → lower cyanogenesis toward urban cores
Red → higher cyanogenesis in surrounding rural populations.

A diagram like this risks making the pattern look more universal than it is. Real data are noisy. Cities are not perfect circles. Parks and rail corridors create fingers of habitat that stretch into the urban core. Rivers and coastlines distort the rings. Clover may be absent from the most paved neighborhoods and abundant in suburbs. Yet the schematic captures the important point: independent cities can generate gradients that point in the same direction.

Why should anyone care whether a lawn plant makes cyanide in one neighborhood and not another? Partly because it tells us that evolution is not confined to wilderness. It is active in places where humans live, work, and raise children. The organisms around us are not static background. They are responding to our management decisions, our pavement, our microclimates, our mowing schedules, and our pollutants.

But there is a more unsettling reason. Parallel evolution implies that some human-made problems reliably produce biological responses. When the response is the loss of a plant defense, we may shrug. When the response is resistance to pesticides, herbicides, or antibiotics, the shrug becomes a bill.

Clover is a gentle example of the same logic. A repeated selection pressure—in this case the blend of urban conditions—can elicit a repeated genetic response. The response can spread quickly because plants produce many offspring and because a trait tied to a few genetic components can shift rapidly in frequency. And it can vary among cities because the selection pressure is not a single dial but a set of interacting dials.

The most honest conclusion is not that cities force evolution into a single predictable groove. It is that cities make evolution legible. They draw sharp environmental boundaries, often over short distances, and those boundaries can carve equally sharp differences into the living fabric of populations. A child picking clover in a park is not merely gathering leaves; they are, without knowing it, sampling a local history of costs and benefits written in chemistry.

5.3 The Toxic Trade-Off

There is a particular kind of insult humans have learned to deliver to other species: a poison they cannot escape.

A forest can be fled, a drought can end, a predator can be avoided by hiding in a new corner of the world. An estuary cannot be abandoned so easily if you are a fish whose entire life is stitched to the tides.

When pollutants enter such places, they do not simply add hardship; they rewrite the rules of survival with a bluntness that makes natural selection unusually easy to see. The shock, for many people, is not that animals die. The shock is that some do not.

In the most contaminated waterways, certain populations persist where they *should* collapse. They spawn, grow, and replenish themselves in sediments that carry decades of industrial history: polychlorinated biphenyls (PCBs), polycyclic aromatic hydrocarbons (PAHs), heavy metals, and a rotating cast of chemicals that were once invisible to regulation and, for a time, invisible even to imagination. The persistence is not a miracle. It is evolution under duress.

Yet evolution rarely hands out gifts without a bill. Resistance to a toxin is not like acquiring a new tool and keeping the rest of your body unchanged. It often requires rerouting fundamental physiology: the way cells sense chemicals, the way enzymes process them, the way development is regulated. Those changes can solve the immediate problem while quietly worsening others. The surviving population becomes, in a sense, specialized for a damaged world.

The most vivid examples come from fish that live where rivers meet the sea. Estuaries collect pollutants the way a doormat collects grit: whatever passes through tends to leave something behind. They are also biologically rich and economically important, supporting fisheries and acting as nurseries for many species. When contamination arrives, the selection pressure can be intense because the toxins are not merely unpleasant; they are lethal to embryos, deforming hearts and jaws, halting development, or preventing hatching altogether.

Along the Atlantic coast of North America, the Atlantic killifish, *Fundulus heteroclitus*, has become an emblem of this kind of rapid adaptation. Killifish are not glamorous. They are small, hardy, and often overlooked, a staple prey item and a common resident of salt marshes. Their unassuming nature is part of why they are such a useful evolutionary protagonist: they are abundant enough to be sampled, widespread enough to provide multiple natural replicates, and tied closely enough to local habitats that populations can differ sharply over short distances.

In several heavily polluted estuaries, killifish populations have evolved striking tolerance to dioxin-like compounds, including certain PCB congeners and related pollutants. These chemicals are dangerous partly because they hijack a deeply conserved cellular signaling pathway. In many vertebrates, a receptor called the aryl hydrocarbon receptor (often abbreviated AHR) binds certain planar aromatic molecules. When activated, it alters gene expression, including genes involved in detoxification. This is, under natural conditions, a sensible system: organisms need a way to sense and process environmental chemicals.

Industrial pollutants exploit that system. Some bind too well or trigger it too strongly, pushing development and metabolism into harmful regimes. Embryos are especially vulnerable because their tissues are growing and patterning at high speed; small disruptions can become catastrophic malformations.

One evolutionary solution is brutally simple: make the receptor less responsive. If the toxin cannot flip the switch, much of the downstream damage is avoided. Genetic studies over the past decade have repeatedly pointed toward modifications of AHR signaling in tolerant killifish populations, including changes involving AHR2, a particular version of the receptor found in many fish. More recent mapping work has suggested that, at least for specific toxic effects such as developmental defects induced by a potent PCB-like compound, resistance can sometimes be explained largely by one or two genetic regions of large effect. That architecture matters because it helps explain speed. If tolerance required tiny shifts at hundreds of loci, the response might be slower or more constrained. When a small number of genetic changes yield a big survival advantage, natural selection can act with startling efficiency.

It is tempting, at this point, to cheer. Nature fights back. Evolution rescues populations. The story seems to offer comfort: life is resilient, and perhaps ecosystems can adapt to whatever humans throw at them.

The comfort is incomplete, because the same modifications that block pollutant toxicity can interfere with normal biology. The AHR pathway did not evolve for the convenience of chemical companies. It is involved in regulating gene expression in ways that interact with development, immunity, and metabolism. Blunting it may be protective in a toxic marsh and costly elsewhere, or costly even within that marsh when other stresses pile on.

The word *trade-off* captures this logic. A trade-off occurs when improving one function necessarily harms another, because organisms have limited resources or because a single biological mechanism influences multiple traits. In evolutionary biology, trade-offs are not moral judgments. They are engineering constraints. If you reinforce a door against burglars by adding heavy locks and thick steel, you may also make it harder to open quickly during a fire.

Toxin resistance can impose costs in at least two broad ways. The first is energetic. Detoxification is expensive. Producing enzymes, maintaining cellular defenses, repairing damage, and excreting modified molecules all require energy and nutrients. If a resistant organism is constantly running its biochemical pumps, it may have less to spend on growth, reproduction, or immune defense.

The second is mechanistic. Resistance often works by altering a sensitive target. In insects, resistance to certain insecticides can evolve through changes in nervous system proteins. Those proteins are not optional; they are central to movement and sensation. Mutations that reduce insecticide binding can also reduce normal function, producing subtle impairments that show up as slower development, reduced fertility, or shorter lifespan. Across many insect groups and many insecticide classes, studies and meta-analyses have found that such fitness costs are common, though far from guaranteed. Sometimes the cost is obvious; sometimes it is small until the environment becomes challenging.

Fish in polluted waters offer a parallel. A killifish embryo that avoids toxin-induced heart defects has cleared the most immediate hurdle: it lives. But altered signaling and detox pathways can change how efficiently it uses oxygen, how it handles temperature stress, or how it copes with low-oxygen conditions that are themselves common in degraded estuaries.

Measurements in some pollutant-tolerant killifish populations have suggested precisely this kind of physiological tax. For example, individuals adapted to PAH-rich environments can show changes in metabolic performance, including reduced ability to ramp up oxygen consumption during demanding conditions. A useful concept here is *aerobic scope*, the difference between the baseline metabolic rate and the maximum metabolic rate. Aerobic scope is the capacity margin that allows an animal to sprint, digest a big meal, escape predators, tolerate heat, or recover from stress. If aerobic scope shrinks, the animal may do fine on a calm day and struggle on a hard one.

These costs can remain invisible until the environment changes. A tolerant population may thrive in a contaminated marsh where the main threat is chemical toxicity, yet become unusually vulnerable when confronted with warming water, hypoxia, or disease. The adaptation is real, but it is a specialization.

There is another fish story, older and more local in its fame, that makes the stakes feel personal. The Hudson River, flowing past one of the most densely populated regions in the United States, became heavily contaminated with PCBs during the twentieth century. PCBs were widely used in electrical equipment because they are chemically stable and resist burning. That stability is exactly what makes them ecologically persistent. They settle into sediments and enter food webs. By the time the scale of contamination was recognized, a great deal of PCB mass had already been distributed along the river system.

The Hudson River tomcod, *Microgadus tomcod*, a small cod-like fish, evolved tolerance in this environment. Research associated with environmental geneticist David Wirgin and colleagues in the 2000s and 2010s helped show that a deletion affecting AHR signaling contributed to resistance. The tomcod case is often told with a double edge: on the one hand, it demonstrates rapid adaptation in a population exposed to a severe pollutant regime; on the other hand, it makes painfully clear what kind of selection was required to produce that adaptation. Enough sensitive individuals had to fail—especially at early life stages—for the resistant genotypes to become common.

A grim arithmetic sits behind every triumphant graph of rising tolerance. Selection does not sculpt without casualties. In a contaminated estuary, the casualties are often embryos that never hatch or larvae that die before anyone sees them. Evolution works in silence, and the silence is full of absence.

Even when a resistant genotype is common, the environment can keep charging interest. Chemical exposures are rarely neat single-molecule challenges. Real sediments hold mixtures: different PCB congeners, PAHs from combustion, metals from industrial runoff, and modern contaminants that interact in nonlinear ways. A genetic change that blocks one mechanism of toxicity may not protect against others. A population can evolve resistance to one dominant threat and still suffer, simply in subtler forms. The marsh may produce fish that survive but grow slowly, reproduce less, or carry developmental scars that make them less competitive.

The logic extends beyond aquatic systems. Consider the way resistance evolves in pest insects exposed to insecticides. Farmers and public health programs, aiming to protect crops and reduce disease transmission, apply intense selection. Insects that carry detoxification enzymes or altered target sites survive. Their descendants inherit the protective variants. Over time, the population shifts. Yet resistant strains often show reduced performance when the chemical is absent: slower larval development, decreased fecundity, reduced mating success, or depleted energetic reserves. Sometimes compensatory evolution reduces those costs later, but the initial trade-off can be substantial.

This is one reason why resistance management strategies sometimes recommend reducing or rotating chemical use. If resistance carries a cost, removing the chemical can allow sensitive genotypes to return because they outcompete resistant ones in a cleaner environment. The key word is *sometimes*. Costs are variable, context-dependent, and can be eroded by further evolution. The same is true in polluted waterways. If a pollutant source is cut off and the environment gradually recovers, the genetic variants that provided resistance may decline, or they may persist if the cost is small or if pollutants remain in sediments for decades.

A simple diagram captures the core idea. Imagine "resistance" on one axis and "general performance" (growth rate, reproductive success, stress tolerance) on the other. In a highly polluted environment, high resistance can correlate with high survival, so it looks beneficial. In a cleaner environment, the relationship can invert: the extra machinery or altered signaling that provides resistance can become dead weight, pulling performance down.

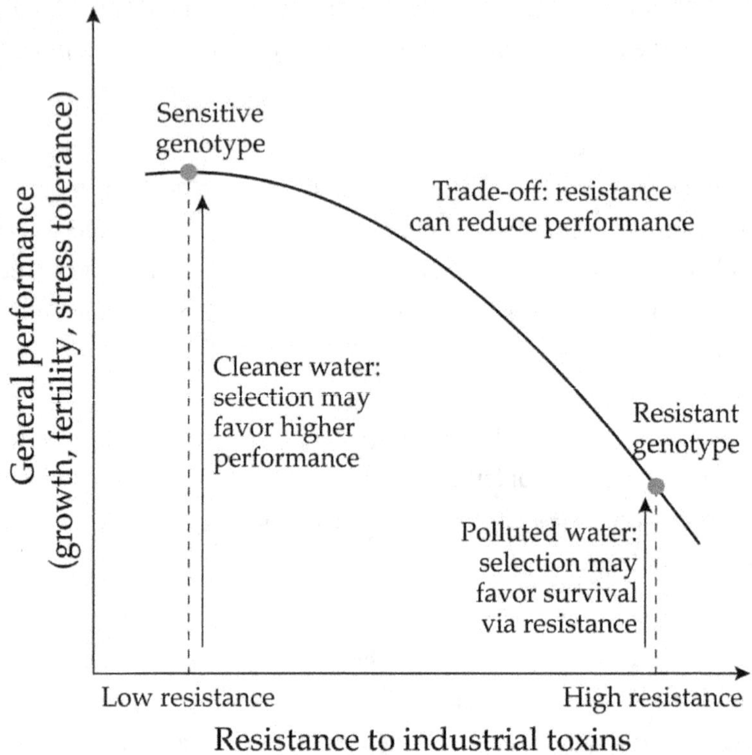

Diagrams tidy the world. Real populations are cloudier. Some resistant genotypes carry little cost. Some costs appear only under heat stress or food limitation. Some are paid not as reduced adult performance but as hidden vulnerabilities in embryos. Yet the conceptual shape is useful because it prevents a common misunderstanding: resistance is not "free immunity." It is a physiological reconfiguration, and reconfiguration has consequences.

The most human aspect of this story is how it scrambles familiar ideas of progress. Cleaner air can allow a moth population to revert toward its pale form. Cleaner sediments, however, are not always quick to arrive. Many industrial toxins persist for decades, locked into mud like a chemical archive. A fish population can remain under selection long after the factory has closed, long after the original discharge has stopped, long after people have stopped thinking about the source. The selection pressure is historical, but the cost is paid in the present tense.

There is also a moral illusion worth dispelling. When a population evolves tolerance, it can look as though the ecosystem has "adapted" in a comforting, self-healing way. What has actually happened is more precise: the population has changed genetically such that individuals that would once have died now survive. That says nothing about whether the broader community has recovered, whether sensitive species have been lost, or whether energy flow through the food web has been altered. Evolutionary rescue is possible, but it can coexist with ecological damage.

And then there is the quietest trade-off of all: a resistant population may become a kind of indicator species for what has been done. The genes that spread are a record of toxins present in the environment. They are, in effect, biological testimony. In a marsh where tolerance has evolved, the fish carry the story of contamination in their genomes and in their physiology. They live, but they are shaped by what threatened to kill them.

On a calm evening in an estuary, the surface can look clean: a few ripples, the smell of salt, a heron standing still at the edge. Beneath that surface, the sediment may hold a chemistry set assembled over generations. A fish that thrives there has not escaped the chemical world; it has been rewritten for it. Survival, in such places, is not simply a triumph. It is a compromise made flesh.

Chapter 6

The Infinite Feast

Humanity's quest to feed billions has triggered a biological insurgency in our fields, where weeds and insects evolve with terrifying speed to dismantle our chemical weapons. By examining the genetic counter-moves of agricultural pests, we reveal the futility of 'shock and awe' agriculture. Only by understanding the evolutionary logic of our enemies can we hope to maintain the harvest without inadvertently breeding invincible super-pests.

6.1 The Chemical Crucible

In the middle of the twentieth century, a new kind of confidence took hold in farming. Insects and weeds had always been there—chewing leaves, stealing light, competing for water—but now they could be *eliminated.* The new chemicals felt like engineering: apply the product, remove the problem, repeat. The trouble is that insects and weeds are not broken machinery. They are populations of living things, full of hidden variety, and chemicals do not simply *kill.* They *select.*

Selection is easy to misunderstand because it looks, at first glance, like brute force. Spray a field and most of the pests die. What remains looks like a lucky remnant. But luck, repeated on an industrial schedule, stops being luck. When the same mode of action is used over and over—the same biochemical "trick," the same molecular target—we turn a farm into a very pure evolutionary experiment. The experiment has one question: *can any individual survive this?* If the answer is ever "yes," even rarely, the experiment immediately turns into a second question: *can those survivors leave more descendants than everyone else?* On a modern farm, the second question is almost always answered "yes," because the chemical has cleared the field of competitors.

A pesticide's "mode of action" is simply the way it harms its target. Some herbicides disable a particular enzyme in a plant's metabolic pathway; some insecticides jam a nerve receptor; some fungicides block a key step in making a cell membrane. These are impressive molecular moves, and they often work beautifully at first. But a single mode of action also creates a single, reliable filter. Put enough living things through a reliable filter, and the population on the other side will look different.

The idea that resistance will appear can sound pessimistic, as if it implies nature is out to get us. It is more like arithmetic. Agricultural fields hold huge populations: millions of weeds seeds in the soil, swarms of insects moving through a landscape, fungal spores drifting like smoke. Within those large numbers sits genetic variation: differences in DNA that arise by mutation, recombination, and the messy history of each species. Most variation does nothing useful. Some is harmful. But a tiny fraction will, by chance, reduce the damage caused by a toxin. When the toxin is applied repeatedly, that fraction is no longer tiny for long.

One reason resistance can feel like a betrayal is that it arrives after success. The first years of a new chemical often look like a triumph, and the market rewards triumph. Farmers adopt what works; companies sell what works; a whole region can converge on the same practice because it is rational, economical, and widely recommended.

The convergence is the evolutionary problem. It turns the landscape into something like a laboratory, except the subjects reproduce faster than the researchers can redesign the experiment.

The pattern is old, and its earliest famous chapters were written indoors. During the Second World War, DDT was celebrated as a miracle insecticide, used against mosquitoes and lice, and later sprayed widely in agriculture and household settings. It helped reduce typhus and malaria risk in some contexts, and it became a symbol of chemical modernity. Yet within a few years of heavy use, resistance appeared in multiple insect species, including houseflies and mosquitoes. The speed mattered: not centuries, not even decades, but a handful of seasons. The details differed by species, but the logic was the same. A population contained individuals with traits that made DDT less lethal—changes in the nervous system that reduced sensitivity, enzymes that broke down the toxin, behaviors that reduced contact. DDT did not create those traits on demand; it rewarded the ones already present, and it rewarded them with the greatest prize evolution can offer: more descendants.

A useful way to picture this is to imagine that a pesticide is a lockpick designed for one particular lock. When it is new, it opens the lock easily. But if you keep using the same lockpick on every lock in town, you should not be surprised that locksmiths start appearing. The locksmiths, in this analogy, are not conscious engineers. They are simply the subset of locks that happen to be slightly different already. Repeated use turns that subset into the new normal.

In weeds, the story is especially stark because weeds have two features that accelerate evolution: sheer numbers and a seed bank. A weed population is not just the plants you see standing between crop rows; it is also the buried archive of seeds from previous years. That archive contains genetic variants from many past seasons, ready to germinate when conditions allow. Apply the same herbicide year after year, and the field becomes a sorting machine that steadily enriches the archive for resistance. Even if a resistant plant is initially rare—one in a million, one in ten million—the scale of agriculture makes "rare" a starting point, not a safeguard.

This is the logic behind what farmers often call the pesticide treadmill. A new product arrives, works extremely well, and allows simplified practices. Over time it becomes less reliable. The response is often to spray more, spray more often, or switch to a chemically similar product that shares the same target. Each of these moves can intensify selection. The treadmill is not just a metaphor for frustration; it is a description of an evolutionary feedback loop in which our attempt to maintain control strengthens the force pushing back.

Herbicide resistance, measured globally, has become too widespread to treat as a local mistake. Ian Heap's continuously updated International Herbicide-Resistant Weed Database reported, as of January 28, 2026, 540 unique resistance cases worldwide when counted as species-by-site-of-action combinations, spanning 273 resistant weed species across 75 countries and 102 crops. Those numbers are not merely a catalogue. They are a map of repeated evolutionary outcomes under repeated human practice.

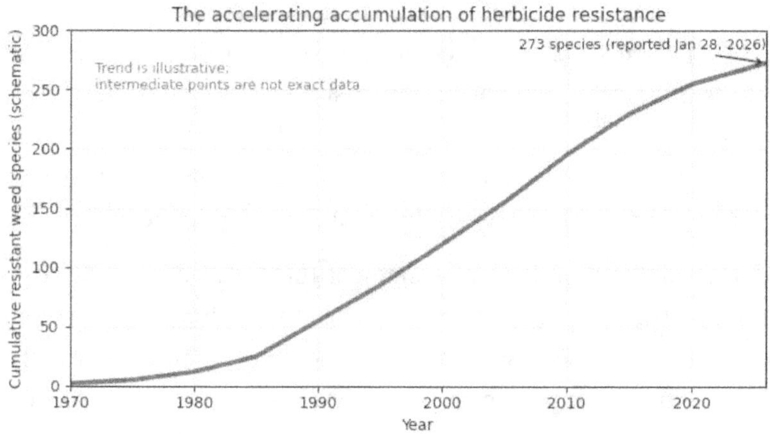

If there is a single chemical that captures the modern tension between simplicity and evolution, it is glyphosate. For years it offered a rare combination: broad-spectrum weed control, relatively favorable toxicological profile compared with some older herbicides, and compatibility with conservation tillage practices that reduce soil erosion. The development of glyphosate-tolerant crops made it even more attractive. A farmer could plant the crop, spray the field, and watch many competitors disappear while the crop stood unharmed. It was efficient, time-saving, and economically compelling. It was also, evolutionarily, a near-perfect pressure cooker.

When selection is intense and consistent, resistance can arise through surprisingly dramatic genetic changes. A vivid example is Palmer amaranth, a fast-growing weed that can overwhelm crops with shocking speed. In some populations, glyphosate resistance did not depend on a tiny tweak to the herbicide's target enzyme. Instead, it involved gene amplification: the plant evolved to carry many extra copies of the gene encoding the enzyme EPSPS (5-enolpyruvylshikimate-3-phosphate synthase), which glyphosate inhibits. More copies meant more enzyme; more enzyme meant the herbicide could not shut the pathway down effectively. Studies in the late 2000s and early 2010s showed that resistant plants could carry dozens of copies of this gene, and that the copy number tracked resistance. What matters for the broader story is not the molecular detail, but the lesson: resistance can be a large-effect trait produced by a large genomic change, and it can spread quickly when a chemical is used ubiquitously.

Gene amplification is also a reminder that resistance is not always a fragile trick. A comforting story sometimes circulates: resistance must come at a cost, so if we stop using the chemical, susceptible types will outcompete the resistant ones and the problem will fade. This can happen, and fitness costs do sometimes slow the spread of resistance. But it is not guaranteed. In Palmer amaranth, evidence has suggested that even very high EPSPS copy number may impose little or no penalty under some competitive conditions. If a resistance trait is not particularly costly, it can persist like an insurance policy in the population. Then the moment the herbicide returns, the insurance pays out.

There are several broad routes by which pests and weeds become less vulnerable to chemicals. One is target-site resistance: a change at the molecular target so the toxin binds poorly, like a key that no longer fits the lock. Another is non-target-site resistance, which is less tidy but often more worrying: the organism reduces uptake, sequesters the chemical away from sensitive tissues, pumps it out, or breaks it down using detoxification enzymes before it can do harm. The phrase "metabolic resistance" refers to this last route. It is, in spirit, like evolving a better liver.

Metabolic resistance is unsettling because it can be promiscuous. A target-site mutation can be very specific: it blocks one herbicide and does nothing else. Metabolic mechanisms can sometimes confer cross-resistance, providing partial protection against several different chemicals, even ones with different modes of action. From an evolutionary perspective, this makes sense. If a landscape is saturated with toxins, any genetic change that improves the general capacity to detoxify is repeatedly rewarded. It is not hard to see the strategic trap: when farmers respond to early resistance by adding more products or mixing chemicals, they may slow certain single-target mutations while simultaneously building a world in which generalist detoxification is the best long-term investment for the weed.

The chemical crucible is not limited to weeds. Insect pests respond just as quickly when the selection is strong and predictable. The same logic plays out in public health, where insecticides used against mosquitoes have repeatedly selected for resistance, complicating malaria control efforts. The overlap in principles is not accidental. Whether the organism is a weed in a soybean field or a mosquito near a village, the evolutionary ingredients are similar: large populations, short generation times, and an environment in which survival is strongly tied to one trait.

One reason single-mode reliance is so potent is that it collapses the complexity of survival into a single exam question. Imagine a university course graded entirely by one multiple-choice test. Students who happen to guess the right answer—or who have seen the exact question before—win disproportionately, regardless of their broader abilities. In a diverse environment, survival is graded on many questions: drought tolerance, competition, predators, pathogens, timing. In a chemically simplified environment, survival can hinge on whether one enzyme works a little differently or one detoxification pathway runs faster. Narrow grading produces narrow adaptation, and narrow adaptation can appear with startling speed.

It is tempting to treat resistance as a property of the pest, as though some species are simply "bad actors." But the deeper pattern is that resistance is often a property of the system we build. Stephen Powles and Qin Yu argued in 2010 that the evolution of herbicide resistance in plants is not surprising when selection is repeated across large populations at a landscape scale. Surprise comes from forgetting how much evolutionary fuel we supply: enormous population sizes, repeated mortality that strongly favors particular variants, and agricultural practices that help survivors reproduce.

Consider what "help survivors reproduce" means in practice. A herbicide does not just kill susceptible weeds; it clears space, light, water, and nutrients. The rare survivor that would otherwise be shaded out now grows in a suddenly generous world. A surviving insect finds not a crowded community but a buffet of plants with fewer competitors for its attention. We sometimes describe pesticides as reducing numbers, but from an evolutionary angle they also restructure opportunity. They turn survival into a bottleneck and then reward those that pass through it with abundance.

The speed of change can be hard to accept because it violates an everyday intuition about evolution as slow. Yet evolution can be fast when selection is harsh and the generation time is short. Many weeds complete a generation in a single season; many insects complete multiple generations in one summer. A decade is not a blink of evolutionary time under these conditions; it can be dozens of rounds of selection. And because farms repeat their practices on a schedule, selection is not sporadic. It is rhythmic, predictable, and geographically extensive.

There is also a subtle psychological effect. When a chemical begins to fail, the first response often frames the problem as insufficient force: *spray earlier, spray later, spray higher, spray twice*. That response is understandable. It treats the field like a battlefield where victory depends on escalation. But escalation is itself information. It tells the population that the toxin will be present, that it will be present often, and that only those who can withstand a large dose will reproduce. In evolutionary terms, that is not just pressure; it is training.

To say this plainly: pesticides are not only weapons. They are also teachers.

They teach through mortality, and mortality is a sharp instructor. If one plant in a million carries a variant that allows it to detoxify a herbicide slightly faster, that plant is irrelevant until the herbicide arrives. Once the herbicide arrives repeatedly, the variant becomes a ticket through the bottleneck. If the same herbicide arrives year after year, the ticket becomes a passport.

This does not mean pesticides are futile. It means they must be understood as part of an ecological and evolutionary relationship, not as a one-time technological fix. The dream of a perfect chemical, forever effective, runs into the hard fact that any consistent attempt to control a reproducing population will favor whatever heritable traits blunt the control. Nature does not need foresight. It only needs survivors.

The chemical crucible is, at bottom, a problem of monotony. A single mode of action used across a whole landscape turns variation into a predictable advantage. The field becomes a stage with one spotlight, and any actor who can stand in that light without burning becomes the star. In such a world, resistance is not an accident. It is the expected outcome of ordinary biology, played at industrial scale.

6.2 Fortresses of Corn and Cotton

By the time genetically modified crops arrived, the logic of resistance had become almost painfully familiar. Chemical pesticides had been strong, consistent filters; the survivors had been the raw material for rapid adaptation. Genetic engineering promised something different: not a new spray, but a new plant. Instead of coating a field with toxin, the crop itself could make the toxin, in the tissues an insect wanted to eat. It sounded, and often felt, like moving from open ground to a fortress wall.

The best-known example is *Bt.*

Bt is short for *Bacillus thuringiensis*, a soil bacterium that produces proteins lethal to certain insects. Long before genetically modified crops, farmers (especially in organic systems) used Bt sprays: formulations of the bacterium or its toxins applied like other insecticides. What genetic engineering changed was delivery. If corn or cotton could be made to produce a Bt toxin in its leaves, stems, or pollen, then the plant would present a built-in hazard to a feeding larva. In many regions, Bt crops reduced insect damage, reduced the need for some broad-spectrum insecticide sprays, and increased yields. For a time, they seemed to break the treadmill.

They did not break evolution.

Bt toxins typically act in the insect gut. A susceptible larva eats plant tissue containing the toxin; the toxin binds to receptors in the gut lining, forms pores, and disrupts the gut so severely that the insect stops feeding and dies. That description is necessarily simplified—biology always is, when compressed—but it captures the crucial point for evolution: there is a molecular interaction that must succeed for the toxin to work. Change the receptor, change the binding, change the downstream steps, and susceptibility can drop. Add to this the fact that insects often have short generation times and huge populations, and it becomes clear why a Bt field is not merely protection. It is selection pressure presented at the scale of a landscape.

The "fortress" metaphor is useful only up to a point. Fortresses have gates. They have weak points. They have attackers who adapt. The strongest wall can become a training ground if it is the only obstacle anyone faces.

A Bt crop is also a peculiar kind of weapon because it is always on.

A farmer can choose when to spray; a Bt plant expresses its toxin continuously across the growing season. Continuous exposure is not automatically bad—it can be extremely effective—but it shapes the kind of evolutionary contest being staged. When exposure is constant, there is less opportunity for susceptible insects to survive by chance. That intensifies selection for whatever genetic variants blunt the toxin's effects. It can also reduce the role of behaviors that sometimes protect insects from sprays, such as hiding during application or feeding at different times. If the plant itself is toxic, the only safe strategy is to be able to tolerate the toxin.

So how can a technology that seems, on paper, to invite rapid resistance remain useful in practice? The answer is that Bt crops were never intended to work as a simple one-step fix. Their durability depends on management that is, at heart, evolutionary.

The most counter-intuitive of these measures is the refuge strategy: leaving some crop area deliberately *non-Bt*, so that susceptible pests can survive.

At first, this can sound like sabotage. Why plant anything that invites damage? Why keep a population of pests alive on purpose? The reason is genetic, and it hinges on a deceptively simple idea: *if most surviving insects are susceptible, resistant insects have trouble finding resistant mates.*

Resistance is often (not always) inherited in a way that makes the refuge strategy powerful. Suppose the resistance trait is recessive. That means an insect needs two copies of a resistance allele to survive well on Bt plants. Let S represent a susceptibility allele and R represent a resistance allele. If resistance is recessive, then only RR insects are strongly resistant; SS are susceptible; and SR heterozygotes are closer to susceptible in their survival on Bt. In that situation, a refuge produces many SS insects. A Bt field produces few survivors, but among those survivors are the rare RR insects. If the refuge is large enough and close enough, RR survivors from the Bt field are likely to mate with SS insects from the refuge. Their offspring are SR, and if SR insects die on Bt plants, the resistance allele is constantly being pulled back into a genetic background that cannot easily express it.

The refuge strategy is, in effect, a way to *dilute* resistance alleles by flooding the mating pool with susceptibility.

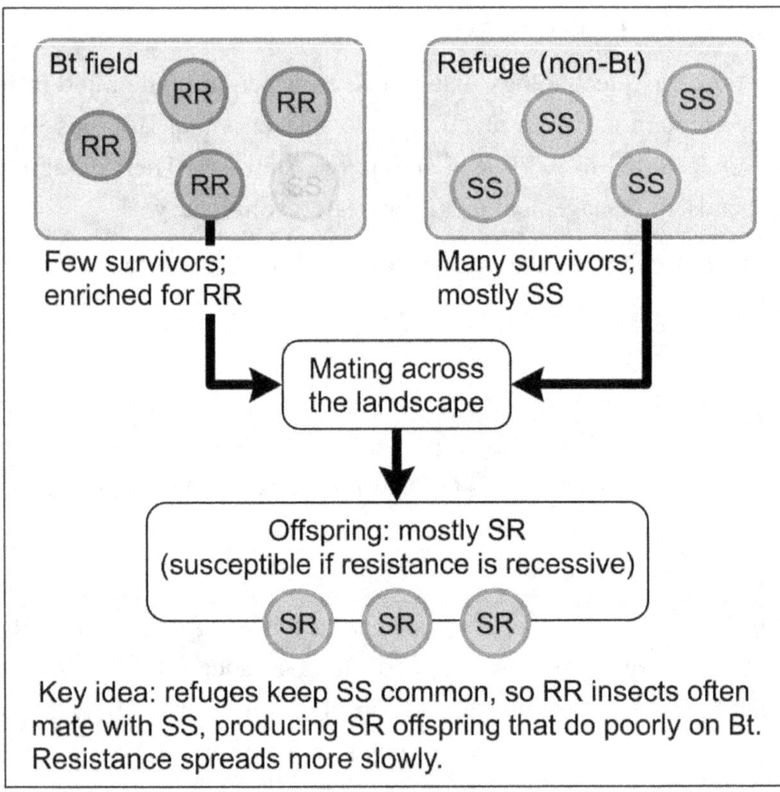

Of course, real genetics can be messier than this tidy cartoon. Resistance is not always recessive. Sometimes heterozygotes survive reasonably well, which weakens the refuge effect. Sometimes there are multiple resistance mechanisms, or multiple toxins in a crop, each with its own genetics. Sometimes insect movement is limited, so refuges do not feed the mating pool of the Bt field effectively. And sometimes, bluntly, refuges are not planted as required, or are planted too late, or are too small, or are treated in ways that kill the very susceptible insects they are meant to preserve.

That last point is uncomfortable because it reminds us that resistance management is not purely a scientific question. It is also a question of incentives, compliance, and trust. The benefits of refuges are partly collective: the region gains durability. The costs of refuges can feel individual: the farmer who plants a refuge may feel they are volunteering for damage while neighbors enjoy full protection. Regulations have tried to formalize the bargain. In the United States, the Environmental Protection Agency has issued insect resistance management requirements for Bt plant-incorporated protectants, specifying refuge sizes and configurations that depend on the crop, the toxin, and the pest group. This is evolution written into policy: a recognition that genes in insects respond to the shape of landscapes humans create.

Another major idea for protecting Bt crops is the pyramid: engineering plants to produce *more than one* Bt toxin aimed at the same pest. If one toxin is a wall, two toxins are meant to be two walls that must be climbed at once. In theory, pyramids can slow resistance because a pest must carry resistance to both toxins to survive, which is rarer than resistance to one. The theory is not magical; it relies on conditions. The toxins must act sufficiently independently, and they must provide what resistance specialists call “redundant killing”: an insect resistant to toxin A should still be killed by toxin B, and vice versa. If the toxins share too much biology—if one resistance mechanism blunts both—the pyramid becomes more like two locks that can be picked with the same tool.

These strategies—refuges and pyramids—sound like refinements, and they are. But they also represent a deeper shift in mindset. Bt crops are not simply products. They are an attempt to design an ecosystem in which resistance evolves slowly enough that agriculture can keep up.

A concrete example of what happens when the design fails comes from the western corn rootworm, often called the "billion-dollar bug" in North American maize. Rootworm larvae feed on corn roots, reducing water and nutrient uptake and making plants prone to lodging. The pest is well matched to modern agriculture because it thrives in extensive corn landscapes and can build large populations. When Bt maize traits targeting rootworm were introduced, they offered a powerful new tool. One of the early traits produced a toxin called Cry3Bb1.

Then, in the field, cracks appeared. By around 2011, research groups reported populations from certain problem fields that survived better on Cry3Bb1 maize than populations from fields without severe injury. Those problem fields often shared a history: the same Bt trait planted repeatedly over consecutive years. In evolutionary terms, this is a familiar recipe. A large pest population is exposed, generation after generation, to a consistent selection pressure. Any genetic variant that increases survival becomes disproportionately represented in the next generation. Over time, the field becomes an arena where the trait is tested, and the pest population gradually shifts from "mostly vulnerable" to "uncomfortably capable."

The rootworm case is revealing for another reason. The original logic of Bt resistance management was built around a "high-dose/refuge" concept: if the Bt crop produces a sufficiently high dose of toxin, it should kill not only susceptible insects but also heterozygotes, leaving only rare homozygous-resistant survivors; refuges then help ensure those survivors mostly mate with susceptible insects. This is a neat, almost elegant plan. But the plan depends on a series of biological assumptions, and rootworm has a way of making assumptions expensive. If the dose is not high enough to kill heterozygotes, or if resistance is not recessive, or if refuges are inadequate, then the genetic arithmetic changes. The dilution that refuges are meant to provide becomes less effective, and resistance can spread faster.

If the chemical crucible taught that monotony accelerates adaptation, Bt crops add a twist: even when the weapon is genetically built into the plant, the same principle holds. Repetition is information, and pests use information.

The success of Bt cotton against some major caterpillar pests also offers a more optimistic story, but one that still has evolution at its center. In several countries, Bt cotton contributed to substantial reductions in insecticide use for certain target pests, and it often improved farmer livelihoods, at least for stretches of time. Where resistance has been delayed, careful management has usually been part of the explanation: refuges that actually function, use of multiple toxins, monitoring that detects early warning signs, and a willingness to treat Bt as one tool among several rather than as a permanent shield. Where resistance has evolved more rapidly, the reasons often echo the rootworm pattern: intense and widespread selection pressure combined with weak dilution of resistance alleles.

The phrase "shift to genetically modified crops" can therefore be misleading. It suggests a swap: spray for seed, chemistry for genetics. In reality, Bt crops are one of the clearest demonstrations that genetics *is* the arena. Whether the toxin is sprayed on leaves or produced in plant cells, the contest plays out in the DNA of pest populations.

This is why resistance management for Bt crops can feel oddly like diplomacy with an enemy that never signs treaties. It requires constraints. It requires cooperation. It requires planning for the world you do not want, because the world you do want—a perfectly defended monoculture—is precisely the world that gives resistance its greatest advantage.

A refuge is a kind of humility rendered in acreage. It admits that total victory is not the goal; durability is. It also admits something more intimate and, for many people, more unsettling: the pests are not merely *there*. They are responding. They are changing, within the span of a farming career, in direct conversation with our strongest interventions.

That conversation can be made slower, quieter, and less costly. But it cannot be ended, because the insects are not listening to our intentions. They are listening to survival.

6.3 Outsmarting the Algorithm

A pest does not plan, but it behaves as if it has read our playbook. When a landscape offers the same crop, the same planting time, and the same control tactic year after year, any gene that fits the pattern is rewarded. The result can feel eerie: weeds that shrug off herbicides that once seemed unbeatable; insects that treat a toxin-laced plant as dinner. The temptation is to answer with escalation—higher doses, more applications, stronger products, more traits stacked into the seed. Yet escalation often tightens the logic of selection. It makes the environment more predictable, and predictability is exactly what evolution exploits.

Integrated Pest Management, usually shortened to IPM, begins from a different premise. It treats pests as part of an ecological system rather than as invaders to be eradicated. That sounds softer than it is. IPM is not a plea for mercy. It is an attempt to manage risk using the one lever that reliably slows rapid evolution: diversity. Diversity of tactics, diversity of timing, diversity of crops, diversity of habitats. Where the chemical crucible and the genetic fortress can turn into single-question exams, IPM tries to turn pest survival back into a complicated final in which no single trick guarantees a pass.

The phrase "integrated" matters. IPM is often misrepresented as simply "use fewer pesticides." Sometimes it does reduce pesticide use, but the deeper point is integration: monitoring, thresholds, and multiple interventions that are chosen because they interact. An insecticide might be part of an IPM program, but it is used with more restraint and more context, not as a reflex. In the best versions of IPM, chemistry is a tool for specific moments, not the atmosphere the field breathes all season.

One of the quiet revolutions in IPM is a shift in what counts as success. In a purely eradication mindset, success is the absence of pests. In IPM, success is keeping pest damage below a level that causes unacceptable economic loss. That idea is captured in the notion of an *economic threshold*: the pest density at which the cost of damage is expected to exceed the cost of control. The threshold is not a moral statement; it is a practical one. It makes room for a crucial ecological ally: natural enemies.

Fields are not empty stages on which pests perform alone. Predatory insects, parasitic wasps, spiders, birds, microbes, and fungi are all capable of suppressing pest populations. Broad-spectrum insecticides can kill these allies as collateral damage, sometimes more efficiently than they kill the target pest. When that happens, the pest rebounds into a world with fewer enemies, and the farmer becomes even more dependent on chemical control. That dependency is not only expensive; it is evolutionarily hazardous. It tends to increase the frequency and intensity of applications, which increases selection for resistance.

IPM can be understood as an attempt to avoid being trapped in that spiral. It asks a series of ordinary questions that, taken together, change the evolutionary landscape. Is the pest actually present, or is this a calendar spray? Is it present at a density that justifies intervention? Is the crop at a vulnerable stage? Is the pest at a vulnerable stage? Can a non-chemical method reduce its impact now, in this field, without creating a new predictable filter?

Those questions are not glamorous. They are also not passive. They require field scouting, traps, degree-day models that predict insect development from temperature, and a willingness to treat the farm as a biological system with feedback. The reward is that interventions become less frequent and less uniform. From evolution's perspective, that is a colder, rougher environment for resistance alleles to spread through.

A historical anecdote helps here, because IPM is not just an academic concept; it grew out of visible failures. In the mid-twentieth century, as synthetic insecticides became widely available, some cropping systems experienced what came to be called pest resurgence and secondary pest outbreaks. A dramatic version of the story played out in cotton. Powerful insecticides knocked down major pests but also wiped out predators and parasitoids. The main target pest might be temporarily suppressed, but other insects that had been minor nuisances suddenly became major problems, freed from their natural enemies. Farmers responded with more spraying, which further destabilized the system. IPM emerged, in part, as a way to escape this trap: to preserve biological control where possible and use chemicals more selectively.

The evolutionary logic of IPM becomes sharpest when we consider resistance management directly. Resistance does not evolve because a chemical is used once; it evolves because the same selection pressure is applied repeatedly, across large populations, with few alternative routes to survival. IPM tries to interrupt at least one of those conditions. It can shrink the portion of the pest population exposed to a given toxin, reduce the number of times the toxin is applied, or make survival depend on multiple traits rather than one.

Rotations are a classic example, and they appear in several forms. Crop rotation changes the environment the pest experiences from year to year: a specialist insect adapted to corn roots does not automatically thrive when the field shifts to soybeans or alfalfa. Herbicide rotation changes the mode of action applied to weeds, ideally so that no single biochemical target is hammered year after year. Insecticide rotation aims to alternate chemical classes so that any resistance allele is not rewarded continuously.

Rotations are often sold with a simple slogan: “switch products and resistance won’t evolve.” Reality is less kind. Rotations can slow resistance, but their effectiveness depends on details that are easy to get wrong. If two insecticides share a resistance mechanism, rotating between them is like alternating two dialects of the same language; a pest that understands one may partially understand the other. If a resistance allele carries little fitness cost, it may persist through the off-years, ready to surge back when the old product returns. If the pest can move across the landscape, it may still encounter the same toxin frequently even if one farm rotates, because neighboring farms do not. The strongest benefit of rotation often comes not from cleverness but from moderation: each chemical is used less often, which reduces selection pressure overall.

Mixtures and “stacks” add another layer of complexity. Intuitively, using multiple toxins at once seems like it should always slow resistance. Sometimes it does, especially when the toxins provide redundant killing and when resistance to one toxin does not help against the other. But mixtures can also create a different kind of evolutionary bargain. When multiple toxins are common in the environment, selection can favor generalist detoxification mechanisms—the metabolic resistance so troubling in weeds—that provide partial protection against many chemistries. The population may evolve away from narrow, target-site tricks and toward broader biochemical defenses. This does not always happen, and it depends on the pest and the toxins involved, but it is a warning against assuming that "more" is automatically better.

This is one reason IPM emphasizes reducing reliance on any single tactic, chemical or otherwise. Reducing selection pressure is not a rhetorical flourish; it is an empirical, recurring recommendation in resistance management because it addresses the core driver. Fewer applications, lower treated area, better timing, and more non-chemical suppression all make the evolutionary payoff of resistance smaller.

An IPM intervention can be mechanical: cultivation, mowing, tillage, flaming, targeted hand removal in high-value crops. It can be cultural: changing planting dates so a crop avoids peak pest pressure, using cover crops to suppress weeds, adjusting irrigation to make conditions less favorable for a disease. It can be biological: releasing parasitoids in controlled systems, conserving predator habitat at field margins, using microbial pesticides. It can be genetic: choosing crop varieties with resistance traits, including Bt, but deploying them with refuges, rotations, and monitoring so that the trait does not become the only line of defense.

The unifying thread is not any particular tool. It is the refusal to be a static target.

A concrete example, drawn from the world of weeds, shows how this works in practice. Consider a farm facing herbicide-resistant weeds in a system dominated by one crop and one herbicide. The immediate impulse is often to add another herbicide or increase the dose. An IPM approach might still include herbicides, but it changes the whole calendar. A cover crop is planted in the off-season, producing a mat of residue that shades the soil and suppresses weed germination. Planting density and row spacing are adjusted so the crop canopy closes faster, depriving weeds of light. Herbicides, if used, are timed to early growth stages when weeds are most vulnerable, and chosen with attention to modes of action and known resistance patterns. Mechanical control might be used at field edges where resistant patches are concentrated, preventing seed set and slowing spread. Harvest weed seed control—methods that capture or destroy weed seeds during harvest—can reduce the number of seeds returned to the soil. Each step alone may seem modest. Together, they alter what it takes for a weed to thrive. Instead of a single biochemical hurdle, the weed faces shade, competition, disrupted reproduction, and only then, perhaps, a carefully chosen chemical challenge.

The same integrated logic can be applied to insects. Scouting identifies whether pest pressure is building. Pheromone traps or light traps provide early warning. Natural enemy populations are observed and protected when possible. When chemical control is necessary, selective products are chosen that spare predators and parasitoids. Sprays are applied when larvae are small and most vulnerable, rather than on a fixed schedule. In Bt systems, refuge compliance is treated not as a bureaucratic nuisance but as part of the biological machinery that keeps the trait effective. Where possible, crop rotation breaks the life cycle of pests tied closely to a particular host plant.

If this sounds like more work, it is. It asks for attention and local knowledge rather than reliance on a single purchased input. It can also be easier than it seems once the habit forms, because the goal is not perfection but stability. A field in which pest populations are suppressed by multiple forces often demands fewer emergency interventions than a field that has been chemically sterilized and then repeatedly reinvaded.

The phrase "outsmarting the algorithm" is a useful way to capture what is really happening. Evolution is not conscious, but it behaves like an optimizer: it amplifies whatever traits increase reproductive success in the environment provided. Monocultures and single-mode controls make that environment simple, so the optimization is fast and the solutions are elegant and infuriating. IPM complicates the environment. It adds noise. It changes the reward structure. It keeps any one trait from being a guaranteed key.

Even small changes in predictability can matter. A pest population that encounters the same insecticide every generation experiences strong, consistent selection. A population that encounters that insecticide only occasionally, sometimes combined with biological suppression, sometimes preceded by cultural practices that reduce population size, and sometimes replaced by a different tactic altogether, faces a weaker and less consistent filter. Resistance alleles can still arise, but their advantage is smaller and their spread slower. Time is bought, and in agriculture time is not an abstraction: time is seasons of workable control, time to develop new tactics, time to avoid catastrophic yield loss, time to keep older tools effective.

The human side of this story is often underappreciated. IPM asks farmers to tolerate a certain amount of imperfection: a few chewed leaves, some weeds at the margins, the sight of an insect that is not immediately killed. That tolerance can be psychologically difficult in a culture trained to equate good farming with clean fields. It can also be economically risky if markets punish cosmetic damage or if labor is scarce. The success of IPM therefore depends on more than biology. It depends on extension services, cooperative area-wide programs, regulations that align individual incentives with collective durability, and consumers who accept that food is grown in living systems.

When IPM works well, it produces a particular kind of calm. The field is not silent; it is busy, full of interactions. Pests appear, but they do not reliably explode. Chemical controls are used, but they do not dominate the story. Natural enemies are present, not as a romantic ideal but as functioning components of a managed ecosystem. The farmer becomes less like a commander ordering annihilation and more like a strategist managing pressure points.

The most honest promise IPM can make is modest: not that resistance will never evolve, but that it does not have to evolve quickly, everywhere, all at once. The agricultural landscape does not have to be a perfectly uniform signal telling pests what to become. It can be a patchwork of obstacles and uncertainties, in which survival demands versatility rather than a single molecular trick. In that patchwork, evolution still happens—it always will—but it has a harder time turning our own consistency into its greatest advantage.

Chapter 7

Time Travel in a Flask

Evolution was once the exclusive domain of historical inference, pieced together from fossils and deep time. Today, by compressing millennia into months within the laboratory and deploying global genomic surveillance, we can witness the mechanics of adaptation as they unfold. We explore the tools that allow us to press 'play' on the evolutionary process to test our theories against hard data.

7.1 The 75,000-Generation Replay

If you wanted to watch natural selection at work, you would not normally choose an animal you can pet, photograph, or name. You would choose something that lives fast, dies quickly, and produces descendants in extravagant numbers. Bacteria are perfect for this: a single cell can become millions overnight, and every new cell is a new roll of the evolutionary dice.

In 1988, Richard Lenski began what sounds like a modest exercise in patience: twelve populations of the same bacterium, *Escherichia coli*, growing in flasks, day after day, in a simple laboratory environment. The aim was not to produce a particular engineered outcome, but to do something closer to watching weather: to let evolution happen under steady conditions and measure its long-term patterns. The experiment became extraordinary for a reason that can be stated in a single sentence but takes a moment to appreciate: almost every day, a sample of each evolving population was frozen and stored, creating a literal archive of its own ancestors.

That one habit turns evolution from a one-way street into a route you can revisit. Evolution in the wild does not preserve its intermediates neatly labelled and retrievable. A species does not keep a museum of its former selves. But a lab freezer can.

The daily rhythm is deceptively simple. Each population grows for about a day in a defined medium, then a small amount is transferred into fresh medium to start the next cycle. This repeated "serial transfer" creates a consistent environment and a consistent bottleneck: most cells are left behind each day, while a small fraction founds the next generation. The bacteria face the same central challenge every cycle: consume limited resources quickly and turn them into descendants. In such a world, any mutation that helps a cell grow a little faster, extract a little more from the food, or survive the day's competitive scramble will tend to spread.

Because bacteria divide so quickly, the timescale is startling. In a human family, "75,000 generations" is not a number that belongs to ordinary imagination; it is older than agriculture, older than cities, older than the first cave paintings. In *E. coli*, it is a few decades of steady work. Every day adds another slice of history.

The freezer makes that history usable. A frozen sample is not like a photograph of the past; it is the past, paused. Thaw a vial, and the ancestors wake up and begin dividing again. This ability to resurrect older forms is why the LTEE is often described, without exaggeration, as a kind of time travel. Not time travel in the science-fiction sense of changing what happened, but time travel in the experimental sense: you can return to a moment, collect a living sample from that moment, and bring it face to face with its own future.

The simplest and most powerful use of this archive is direct competition. Suppose you want to know whether the bacteria have adapted. You could measure growth rates, cell sizes, or the chemistry of the medium, but these can be misleading. Fitness, in evolution, means something plain: leaving more descendants than your rivals in the same environment. In the LTEE, this can be measured by mixing an evolved strain with its ancestor and letting them compete in the same flask. If the evolved strain increases in frequency, it is fitter. If it declines, it is not. Because the ancestor can be revived from the freezer, you can always compare today's bacteria to the same baseline, even decades later. You can also compare bacteria from any two timepoints, as if you had arranged a reunion between distant relatives who could never have met.

When this is done across thousands and then tens of thousands of generations, a pattern appears that is both intuitive and oddly poignant. Adaptation comes quickly at first. The early generations contain many easy opportunities: mutations that fix obvious inefficiencies, exploit the available nutrients better, or tune basic physiology to the flask environment. But the gains do not continue at the same pace forever. Over time, fitness keeps increasing, yet the curve bends; improvement becomes harder. This is the experimental face of what evolutionary biologists call *diminishing returns*: as a lineage becomes better adapted to a particular setting, the remaining beneficial changes tend to be smaller, rarer, or more constrained by trade-offs.

It is tempting to imagine that diminishing returns must lead to a plateau, as if evolution were climbing a hill with a clearly defined summit. The LTEE complicates that picture. Fitness increases slow down, but they do not simply stop. The bacteria keep finding new ways to be a little better than they were before, even in a seemingly unchanging world. Evolution does not have to reach a final form to remain busy; it only needs the continuing possibility of small advantages.

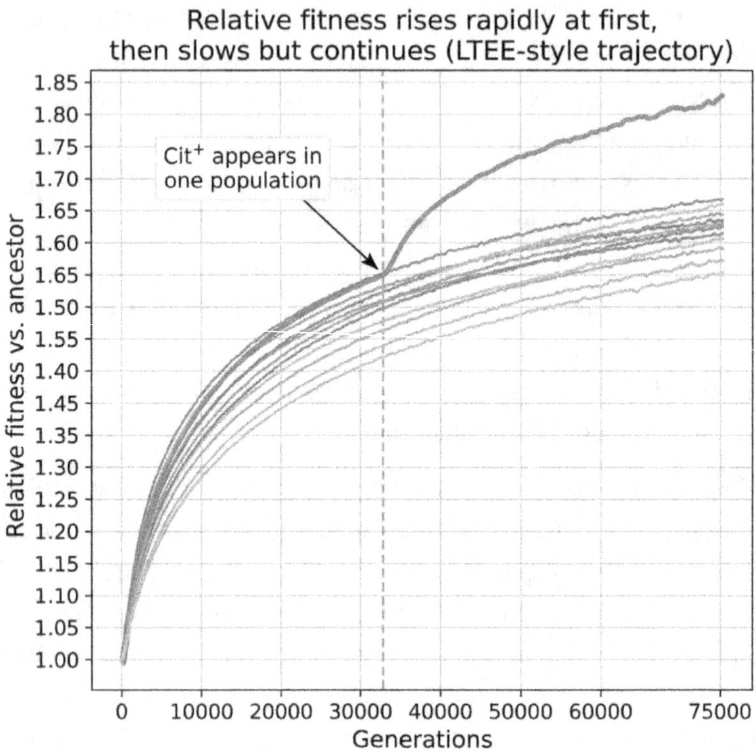

The curve matters, but the freezer matters more, because it lets that curve become more than a description. It becomes something you can interrogate. If fitness gains slow down, why? Is it because beneficial mutations become rarer, because they become smaller, or because beneficial mutations interfere with each other? In asexual populations like *E. coli* (which do not regularly exchange genes the way many plants and animals do), competition can occur not only between individuals but between entire genetic lineages. Two different beneficial mutations may arise in different cells, and because the genome is inherited as a package, those lineages must compete;

only one can eventually dominate. This phenomenon, called *clonal interference*, means natural selection does not simply pick the single best mutation out of a menu. It runs a tournament of whole lineages, each carrying its own bundle of changes.

There is a second, subtler reason the archive is invaluable: it helps separate what is typical from what is contingent. People often ask whether evolution is predictable. If you "replayed the tape of life," as the paleontologist Stephen Jay Gould famously put it, would you get the same outcomes? The LTEE does not replay all of Earth's history, but it does something unusually close to an experiment on that question. There are twelve populations, started from the same ancestor, living in the same conditions, evolving independently. Some outcomes happen again and again across the flasks, suggesting that natural selection repeatedly finds similar solutions. Other outcomes happen once, or not at all, hinting that chance events and historical sequence can matter.

A vivid example is the trait that made the LTEE famous outside evolutionary biology: the ability of one population to use citrate as a food source in the presence of oxygen. Citrate is familiar from oranges and lemons, but in microbiology it is also a standard ingredient in growth media. Many bacteria can metabolize citrate. The ancestral *E. coli* used in the LTEE could not do so aerobically, which is one reason citrate was included in the medium: it was present, but effectively unavailable, like a locked pantry.

After more than 30,000 generations, in one and only one of the twelve populations, bacteria arose that could tap that pantry. The innovation is often abbreviated Cit^{+} (for citrate positive). It was a striking event for several reasons. First, it seemed like a leap into a new niche inside the same flask. Second, it was dramatic in its consequences: once citrate became usable, the population could grow to much higher densities, altering the ecological landscape of the flask. Third, it appeared after a very long wait, which immediately raised the question: was this just a fantastically rare mutation that happened to occur once, or did something about that population's earlier history make the innovation possible?

The freezer archive allowed an answer that goes beyond storytelling. In 2008, Zachary Blount and colleagues, working in Lenski's laboratory, performed what are sometimes called *replay experiments*. The logic is almost disarmingly direct. Take frozen samples from the population that eventually evolved Cit^+, but take them from different times: early in the experiment, intermediate, later, close to the moment before citrate use appeared. Start each sample anew under the same conditions and see whether Cit^+ arises again. If citrate use is simply an ultra-rare mutation, it should be roughly equally unlikely at all times, except for the gradual increase in total number of cells produced. If, instead, earlier mutations "set the stage," then the probability of evolving Cit^+ should change over time.

What they found revealed historical contingency in a clean experimental form. Cit^+ was not equally accessible at all times. Early ancestors almost never produced it, even when enormous numbers of cells were screened. Later ancestors produced it more readily. In other words, the lineage's past had altered its future evolutionary possibilities. Certain mutations that were themselves beneficial (or at least not harmful) had accumulated first, and these changes *potentiated* the later innovation: they made it possible for a subsequent mutation to have the right effect in the right genetic context.

In 2012, Blount, Jeffrey Barrick, and colleagues went further by using genomic analysis and genetic reconstruction to piece together a stepwise path. The details are intricate, but the core idea is easy to grasp if you think of a trait like a machine assembled from available parts. A single mutation rarely invents a complex new function from nothing. More often, evolution reuses and repurposes existing components, altering when and where genes are expressed, duplicating segments, or rewiring regulation. The citrate story showed a plausible sequence: a background of potentiating changes, then an "actualizing" mutation that produced a weak form of citrate use, followed by "refining" changes that improved it. The freezing archive made it possible to identify not only what the final bacteria could do, but how that ability emerged, and to test the causal role of intermediate steps rather than simply narrating them.

There is a temptation, when hearing about a rare innovation, to treat it as a fluke: lightning in a bottle, interesting but not representative. The LTEE invites a different view. Rare events are part of evolution's repertoire, but their rarity can itself be structured. The probability of an innovation may depend on the order in which earlier, more ordinary adaptations occurred. In this sense, "chance" is not a single thing. There is chance in which mutations arise on a given day, and there is chance in which history a population happens to build, a history that can open some doors while quietly closing others.

The multi-population design also makes it harder to fool ourselves. If a single flask evolves a striking trait, it could be a peculiarity. If similar genetic targets are hit across many flasks, again and again, that suggests strong and repeatable selection pressures. One of the recurring themes across long-term bacterial evolution is *parallel evolution*: different populations acquire different mutations, but often in the same genes or pathways, implying that there are only so many effective ways to solve the challenges posed by a particular environment. Parallel evolution is not the same as perfect predictability, but it is a kind of predictability that sits comfortably alongside contingency. Evolution can be repeatable in its broad strokes while still being creative and idiosyncratic in the details.

The freezer archive makes these patterns measurable rather than philosophical. It lets researchers answer questions that would otherwise be speculation. Did a particular mutation increase fitness when it arose, or was it a passenger that happened to ride along? You can reconstruct it in an ancestral background and compete it against the unmodified ancestor. Did the same genetic change have the same effect later in evolution? You can insert it into an evolved background and measure the difference, revealing a common phenomenon known as *epistasis*, where the effect of a mutation depends on the rest of the genome. Diminishing returns can be understood this way: the same beneficial change can be strongly advantageous early on and only mildly advantageous later, because earlier adaptations have already captured much of the low-hanging fruit.

Even the practical risks of a decades-long experiment become part of the story. Laboratories are human places. Freezers fail, labels fade, people graduate, equipment changes, and small mistakes can accumulate like dust. The LTEE's design treats those risks as expected rather than shameful. The archive means a population can be restored from a recent frozen sample if something goes wrong. It also means suspicious results can be audited: if a lineage behaves oddly, you can check whether the oddity is repeatable from an earlier frozen vial or whether it arose from contamination or mix-up. The freezer does not merely preserve history for scientific romance; it makes the experiment robust against the ordinary fragility of long time spans.

By now, the LTEE has become a kind of common language for thinking about fast evolution. When people worry about antibiotic resistance spreading through a hospital, or pests adapting to pesticides in a field, they are thinking about evolution on a human timescale, unfolding quickly enough to matter within careers and lifetimes. The LTEE does not model those exact environments, but it shows what relentless selection can do even when conditions are steady and resources are simple. It shows that adaptation can be rapid, that improvements can keep accruing for tens of thousands of generations, and that a lineage can stumble upon an unexpected innovation after years of apparently routine change.

Most of all, it shows what changes when you can revive the past. Without a frozen archive, "ancestor" is a metaphor. With an archive, the ancestor is a living competitor you can pipette into a flask. The gap between history and experiment narrows until it almost disappears. A bacterium from 1988 can meet its descendant from 2026 not as a reconstructed story, but as two swarms of cells jostling for the same molecules, in the same glass, under the same light, settling the question of "who is fitter?" in the only currency evolution ever really counts: offspring.

7.2 Reading the Source Code

The freezer gives you ancestors on demand, but it does not tell you *how* they became different. Two bacterial strains can grow at different speeds for reasons that look, from the outside, almost identical: one might import sugar a little faster; another might waste less energy on a pathway it no longer needs; a third might have altered regulation so that the right genes turn on at the right moment in the daily cycle. For a long time, experimental evolution relied on such outward signs. Colonies looked rough or smooth. Cells were larger or smaller. A culture turned cloudy sooner. These are real biological changes, and they can be measured carefully, yet they are like watching a city from an airplane at night: you see patterns of light, not the wiring underneath.

High-throughput DNA sequencing changed that. It made it possible to treat evolution as something you can read, line by line, in the organism's genetic "text". In bacteria, the "text" is a genome of a few million letters. Each cell copy is almost the same as the last, but not perfectly; mutations introduce differences. Natural selection then acts on those differences, amplifying some and removing others. If the freezer archive is a living library of earlier editions, sequencing is the ability to compare the editions directly, and to notice exactly which words changed, which paragraphs were duplicated, and which pages were torn out.

This shift did not happen because biologists suddenly became obsessed with DNA for its own sake. It happened because the old questions became answerable in a new way. When fitness rises quickly and then slows, is it because beneficial mutations are disappearing, or because they are being crowded out by interference among lineages? When one population evolves citrate use and the others do not, is the difference a single lucky accident, or a sequence of earlier changes that quietly opened the door? When a population seems stable, is it truly sitting still, or is it seething with genetic competition beneath an unchanging surface?

Sequencing also changed the scale at which it was sensible to ask questions. In the early days, a lab might spend months characterizing a single mutant. Now, hundreds of genomes can be read as part of a single study, and tens of thousands of genomes can be analyzed across global pathogen surveillance. The technique is not merely a better microscope. It is a new form of record-keeping.

A good way to feel the difference is to imagine a familiar scene: a doctor prescribes an antibiotic for an infection, the patient improves, and everyone hopes the story ends there. Sometimes it does not. The bacteria rebound, the infection returns, and now the antibiotic that worked last time has become useless. In the past, this looked like a baffling change in behaviour: the microbes had "learned" something. With sequencing, the story becomes precise. A particular mutation has altered a drug target, or increased the activity of an efflux pump that expels the drug, or changed membrane permeability. If you sequence samples over time, you can watch that mutation rise in frequency within the bacterial population. You see the genetic signature of selection directly.

The same logic applies in the controlled, simplified world of the Long-Term Evolution Experiment. The difference is that the LTEE gives you an unusually clean baseline and an unusually deep history. As the experiment ran into tens of thousands of generations, researchers began sequencing clones (single bacterial isolates) from different populations at different timepoints. Each clone is a snapshot: one genome pulled from the swarm. Compare a clone from generation 10,000 to the ancestor, and you see dozens of differences. Compare generation 50,000 to the ancestor, and you see more. Some differences are single-letter substitutions. Some are small insertions and deletions. Some are large rearrangements. Some involve mobile genetic elements that can copy and paste themselves around the genome. From a distance, "mutation" sounds like a uniform category, but in bacteria it is more like a toolbox of ways to change.

At first, it is natural to imagine that if you list all the genetic differences between ancestor and descendant, you have explained evolution. You have not. Many mutations are passengers. They happen to occur in the lineage that carries a beneficial change, and because bacteria reproduce clonally, the passengers hitchhike along. Other mutations may have been beneficial at the moment they arose but later became irrelevant once other changes appeared. Some may even be harmful, tolerated only because selection on other sites was strong. A genome sequence is a record of what happened, not a guarantee of why it mattered.

The key trick is to combine sequencing with the experimental advantages already built into the LTEE: freezing, replay, and direct competition. When sequencing suggests that a certain gene has changed again and again across independent populations, you have a clue that it matters. When sequencing shows a mutation appearing and then sweeping to fixation (meaning it becomes present in essentially every cell), you have a clue that selection favoured it. When sequencing reveals that multiple lineages coexist for long periods, you have a clue that ecology is unfolding within the flask, or that clonal interference is preventing any one lineage from quickly taking over. But the decisive step is often to *test* a candidate mutation: to reconstruct it in an earlier genetic background, or to compete strains that differ only in that change. The genome gives hypotheses; the freezer and the flask let you interrogate them.

The moment this became widely practical is itself a small historical anecdote about the culture of science. In the early 2000s, genome sequencing was still expensive enough that it could feel like a special event: a big purchase, a single organism, a celebratory paper. Then sequencing costs collapsed, and "resequencing" (reading the genome of something you already have a reference for) became routine. In experimental evolution, this created a new genre of work: time series genomics. Instead of sequencing one endpoint, you could sequence many points along the trajectory. Suddenly, evolution looked less like "before and after" and more like a film.

To make that film useful, researchers also needed ways to translate raw sequencing reads into interpretable mutations. Tools like *breseq*, developed in Jeffrey Barrick's lab (2014), were built for exactly this situation: a haploid microbe, an ancestor genome as reference, and evolved samples that may contain single-nucleotide variants, small indels, deletions, and structural changes. The achievement was not only computational. It was epistemic: it made a complicated process legible enough that an evolutionary biologist could look at a list of candidate mutations and begin thinking mechanistically again, rather than drowning in data.

Once you can read genomes routinely, a new kind of question becomes unavoidable: how much of evolution is visible if you stop focusing on the fate of a single clone and start focusing on the population as a whole?

When you isolate a single colony and sequence it, you are reading one thread of the story, not the whole tapestry. In a bacterial population, especially a large one, many beneficial mutations can arise in different individuals around the same time. Those lineages then compete. Some go extinct. Some persist for thousands of generations before losing. Some briefly dominate and then are displaced. The outward phenotype may not change much while this genetic drama plays out. It is like a political landscape in which parties rise and fall while the country's skyline looks the same to a casual visitor.

Whole-population sequencing (often called metagenomic sequencing in this context, though the term is broader) makes those hidden contests visible. Instead of isolating a single clone, you sequence DNA from the entire population sample. What you get is not a single genome but a set of allele frequencies: at a given position in the genome, what fraction of reads carry letter A, and what fraction carry letter G? Track that fraction over time and you can watch a mutation's rise, stall, decline, or sweep.

Benjamin Good and colleagues (2017) used dense temporal sampling to follow these dynamics in the LTEE. One of the striking outcomes of such work is how busy evolution remains even when overall fitness gains slow. Alleles rise and fall, multiple lineages coexist, and selective sweeps can be incomplete or interrupted. The population is not marching in lockstep toward a single best form. It is a crowded marketplace of alternatives, with natural selection acting on differences that are often too subtle to see without the genome as a readout.

This population view also reframes what "repeatability" means. If you only look at endpoints, you might conclude that evolution is fairly predictable because different populations often end up with improved fitness and similar broad traits. But if you watch the frequency trajectories, you see that the route to those endpoints can vary widely. Some populations experience dramatic sweeps. Others show long periods of coexistence. Some lineages rise on one mutation and later lose to another lineage with a different set of changes. Predictability becomes a matter of scale: the overall direction is repeatable, but the detailed path is contingent.

Sequencing also made it possible to treat mutation rate itself as an evolving trait rather than a fixed background parameter. It is easy to say "mutations happen," and to leave it at that. But organisms can differ in how often mutations occur, and in bacteria those differences can evolve rapidly. A lineage that mutates more frequently is more likely to generate a beneficial mutation simply because it is buying more lottery tickets. The cost is that it also generates more harmful mutations, creating a genetic burden known as *load*. Whether a higher mutation rate is favoured depends on the balance between these forces, and that balance can shift over time.

In the LTEE, some populations evolved "mutator" phenotypes, often through defects in DNA repair genes. Sequencing provided a clean signature: these lineages accumulate mutations at a much faster rate, including many changes that appear effectively neutral, such as synonymous substitutions (DNA changes that do not alter the amino acid sequence of a protein). Work by Sébastien Wielgoss and colleagues (2012) documented not only the rise of strong mutators but also the later appearance of "anti-mutator" changes that partially reduced mutation rates. The story is almost economical. Early in adaptation, when there are many beneficial mutations available, a high mutation rate can be advantageous because it accelerates discovery. Later on, when beneficial opportunities are scarcer and the genome is already well-tuned to the environment, the same high mutation rate becomes more costly because it mostly produces harmful changes. Sequencing turned this from a clever theoretical argument into something you could observe: a tension written into the genome's accumulating scars.

The LTEE's most famous innovation, citrate use, became a proving ground for this new genomic mode of explanation. The trait itself could be detected by growth tests. The drama of its appearance could be charted in population density. But the real conceptual leap was to show that an innovation could be dissected into genetic steps, and that those steps could be tied to changes in evolutionary probability.

When Blount, Barrick, and colleagues combined genome sequencing with reconstruction logic (2012), they articulated a useful vocabulary for thinking about complex traits: *potentiation*, *actualization*, and *refinement*. The words are not magic; they are labels for a pattern that appears in many evolutionary contexts. Potentiation refers to earlier changes that make a later innovation possible or more likely. Actualization refers to the first mutation (or set of mutations) that produces the new trait in some rudimentary form. Refinement refers to subsequent changes that improve the trait's performance and integration into the organism's physiology.

The importance of this scheme is not merely that it tells a satisfying story about one population in one experiment. It offers a way to talk about contingency without mysticism. "History matters" can sound like a vague slogan. Genomics makes it testable: particular earlier mutations alter the effect of later mutations. The genome becomes a record of constraints and opportunities.

There is a further, sobering implication. If key innovations depend on potentiating backgrounds, then predicting evolution requires more than knowing which mutations are possible in principle. It requires knowing which genetic contexts are likely to arise under a given set of selective pressures. In medicine, this is not an academic point. The evolution of antibiotic resistance can depend on the order in which drugs are used, because earlier resistance mechanisms can change the fitness landscape for later ones. The same is true for viral evolution under immune pressure: certain mutations open routes that were previously inaccessible, while others close them. Sequencing does not automatically make evolution predictable, but it makes the sources of unpredictability identifiable.

The phrase "reading the source code" invites a comparison with software, and the comparison is useful if handled carefully. A genome is not a human-designed program. It has no foresight, no documentation, and no guarantee that similar "functions" are implemented similarly across lineages. But there is a real analogy in the way small changes can have large consequences, and in the way the effect of a change depends on context. In a large codebase, altering a line can fix a bug, introduce a new feature, or break something elsewhere. In a genome, a point mutation can tune a protein's function, change gene regulation, or destabilize a system in ways that only become obvious under particular conditions. The LTEE showed that even in a simple environment, a bacterial genome is not a static blueprint but a responsive, evolving system, where selection continually edits what works.

Sequencing also altered the emotional texture of experimental evolution. There is a particular kind of astonishment in realizing that a population you can hold in a small tube contains, at any moment, a multitude of genetic alternatives, each one a possible future. The freezer archive stores those futures in a way that feels almost unfair: you can go back and ask what would have happened if a rare lineage had not gone extinct, or if a potentiating mutation had not occurred, or if a mutator had arisen earlier. You cannot do that with dinosaurs, or with the origin of mammals, or with the evolution of the human brain. But you can do it with bacteria.

And once you begin to see evolution as a stream of genetic edits, a final lesson becomes hard to ignore: selection is not only about spectacular innovations. It is also about innumerable small improvements that never announce themselves as "new traits" at all. A regulatory tweak that saves a little energy. A deletion that removes an unnecessary function. A membrane change that slightly improves nutrient uptake. In the daily life of a bacterium in a flask, these are the differences that decide who leaves more descendants by tomorrow morning.

Sequencing makes those differences visible, and visibility changes what counts as an explanation. The bacteria are still doing what they have always done: dividing, competing, dying, and occasionally stumbling on a mutation that shifts their prospects. The novelty lies in the fact that we can now watch the genetic bookkeeping of those events as they unfold, with enough resolution to see not just that evolution happened, but how relentlessly it keeps rewriting even the simplest forms of life.

7.3 The Crowd-Sourced Laboratory

A laboratory bench is a place of control. You choose the organism, the temperature, the food, the timing, the measurements. You label your tubes and you keep your freezer inventory tidy. It is hard to overstate how powerful that control is for understanding evolution, because it allows clean cause and effect. Yet control has a cost: scale. A single lab cannot watch every city block, every roadside verge, every harbour wall, every backyard bird feeder, every drainage ditch where mosquitoes breed, every patch of herbicide-sprayed weeds. Much of fast evolution happens in precisely those messy places, driven by human choices that vary street by street and season by season.

Citizen science changes what is possible. It takes the apparatus of observation—the eyes, phones, curiosity, and local knowledge of non-specialists—and turns it into a distributed sensor network. The organisms are not brought into the lab; the observers go out into the world. For studying adaptation on human timescales, this is not a charming add-on. It is often the only way to measure evolution at the scale at which people are driving it.

The basic idea is older than smartphones. For centuries, naturalists have relied on letters, specimen exchanges, and local collectors. Charles Darwin's work was built not only on his own observations but on correspondence with pigeon fanciers, gardeners, ship captains, and colonial administrators. The difference today is that the same impulse has been coupled to a camera, a GPS receiver, and an online platform that can store millions of records, time-stamp them, map them, and share them instantly. This changes the labour of science. It also changes the kinds of evolutionary questions that can be asked.

Consider what it means to claim that a trait is evolving in response to cities. A city is not one environment but many: warmer "heat islands," saltier roadside soils, fragmented habitats, altered light cycles, polluted air, novel predators and prey, and constant human disturbance. Urban environments are also replicated across the planet. The centre of Boston is not the centre of Berlin, but both share patterns of pavement, traffic, and microclimate that rural forests do not. If evolution is responding to those shared pressures, you might expect *parallel evolution*: similar traits increasing in similar ways in different cities. Demonstrating that parallelism requires data from many cities, collected in similar ways, over wide geographic ranges. That is exactly the kind of task that overwhelms a typical research group but becomes feasible when observation is shared.

The platforms matter. An app like iNaturalist does more than host photographs. It imposes a structure: an observation is a picture, a place, a date, and a suggested identification. Other users can confirm or correct the identification, and the system can flag records that reach a standard of reliability. The result is not perfect data, but it is data with a provenance and a community of reviewers. Just as importantly, it creates an incentive loop: contributors receive feedback, status, and the satisfaction of being useful. That social infrastructure is as important as the technical infrastructure.

To see why this can illuminate evolution, it helps to focus on a trait that is both biologically meaningful and easy to measure in the field. White clover, *Trifolium repens*, is an unassuming plant that most people recognize as lawn "weed" with three-leaved leaflets and occasional four-leaf surprises. Some white clover plants can produce hydrogen cyanide when their tissues are damaged, a chemical defence that can deter herbivores. This "cyanogenesis" is controlled by genetic variation that can be scored with simple assays. It also carries trade-offs: producing defensive compounds can be costly, and the underlying genetics can interact with temperature and frost risk. In many regions, the frequency of cyanogenic plants differs between warmer and colder environments.

Urban environments add a twist. Cities tend to be warmer than surrounding rural areas, but they can also create winter hazards through frost exposure in open, disturbed patches, and they can impose novel patterns of mowing and herbivory. When researchers ask how cyanogenesis varies from city centres out into rural surroundings, they are often looking for a gradient: does the frequency of the defensive trait change systematically with urbanization? Answering that in one city is interesting. Answering it across many cities starts to look like a general evolutionary phenomenon.

This is where citizen science becomes less about collecting curiosities and more about testing evolutionary ideas. If volunteers sample clover across neighbourhoods and park edges, recording location carefully, the dataset can quickly become large enough to reveal patterns that would otherwise be hidden in noise. One person's sample is anecdote. Hundreds of people sampling in a standardized way is a map of natural selection's footprint.

Urban evolution offers many such traits. Some are visually obvious, like the colour patterns of moths or the banding of snails. Some are behavioural, like birds adjusting their songs in noisy environments. Some are physiological, like plants tolerating heavy metals. The common feature is that the relevant environmental variation is not confined to one plot of land. It sprawls.

Citizen science also changes how quickly you can detect change. Traditional ecological studies often revisit the same sites annually or seasonally, constrained by the field season and the size of the research team. With a distributed network, observations can be continuous. A sudden bloom, an early emergence, a range expansion, a mass mortality event: these become detectable not because a scientist happened to be there, but because someone was there.

Evolutionary change can be rapid enough that this matters. A pesticide is introduced, and within a few years resistant insects become common. A new pathogen enters a region, and hosts evolve different levels of susceptibility. A predator arrives, and prey shift their behaviour. In such cases, time is not a background variable; it is the thing you are trying to measure. Dense observation in time, not only in space, becomes crucial.

The genome-sequencing revolution adds another layer. It might seem that citizen science is about photographs and field notes, while genomics is about machines in specialist facilities. In practice, the two are increasingly connected through shared data ecosystems. A vivid example comes from pandemic-era genomic surveillance, where the "crowd" is not only individuals submitting photos of butterflies but laboratories around the world uploading viral genomes, and teams building tools that turn those genomes into interpretable, time-stamped maps of spread and mutation.

The platform Nextstrain, developed by James Hadfield and colleagues (2018), made a new kind of public scientific object: a continuously updated view of pathogen evolution that integrates genetic changes with time and geography. It allowed viewers—scientists, public health officials, journalists, curious citizens—to see viral lineages appear, diversify, and move across regions. The power of such a system depends on open sharing, frequent updates, and a willingness to treat genomic data as a common resource rather than a private trophy. It is, in a sense, citizen science at the institutional scale: a crowd of contributors generating a dataset too large and too dynamic for any single group to own.

This style of surveillance also exposed the friction points. Sharing data quickly raises questions of credit, priority, and national sovereignty. GISAID, one of the major platforms for sharing influenza and coronavirus sequences, developed a model that tries to encourage rapid sharing while preserving acknowledgement for data generators through its data access agreement. The social design is again as important as the technical design. A sequence is not just information; it is work, and work happens within institutions that respond to incentives and recognition.

The computational challenges of crowd-scale genomics are not trivial. When a pandemic produces millions of sequences, building a new phylogenetic tree from scratch each time becomes painfully slow. Methods such as UShER (Ultrafast Sample placement on Existing tRee) were developed to place new sequences onto an existing large tree rapidly, using mutation-annotated structures and efficient placement algorithms. The point is not the name or the acronym; it is the shift in what "real time" means. A tool like UShER turns an overwhelming flood of genomes into something that can still be navigated as a living map, rather than a static archive.

A further step brought genomic surveillance even closer to everyday life, and in a strange way returned it to the logic of a flask: wastewater sequencing. Sewage is an unglamorous sample, but it is also a pooled biospecimen from a community. If a virus is circulating, fragments of its RNA can end up in wastewater. By sequencing those fragments, it becomes possible to estimate which viral lineages are present in a city, sometimes even before clinical testing catches up. Tools such as Freyja were developed to "deconvolute" mixed wastewater samples, estimating the proportions of different lineages based on shared mutation barcodes.

This is crowd-sourced biology in a different sense. No one is deliberately donating a sample for viral sequencing when they flush a toilet, yet the resulting data can inform public health in a way that individual testing cannot. It is a reminder that "citizen science" need not always mean active participation. It can also mean that the traces of collective life become measurable, and that measurement can guide decisions that feed back into selection pressures on pathogens. Vaccines, antivirals, masking policies, travel patterns, and immune landscapes all alter which viral variants thrive. Genomic surveillance does not merely observe evolution; it can shape the conditions under which evolution proceeds.

There are important limitations, and they are not merely technical footnotes. Wastewater RNA is often fragmented. Viral abundance can be very low when incidence is low, making detection unreliable. PCR primer schemes can drop out certain genome regions, biasing the apparent frequencies of variants. Phylogenetic placement becomes uncertain when data are ambiguous or incomplete. Barcode-based deconvolution depends on existing lineage definitions, which can lag behind reality when new variants emerge. Each limitation is a reminder that reading evolution is not the same as controlling it. The world is not obliged to produce clean samples.

The most successful citizen science projects in evolution acknowledge this messiness and build methods that are robust to it. They also choose questions where imperfect data can still be informative. A photograph may not reveal the genotype of an insect, but it can reveal colour pattern. A GPS-tagged plant observation may not reveal the chemistry of its leaves, but it can reveal range limits and phenology. A wastewater sample may not yield a perfect genome, but it can reveal the presence of a lineage. The art is to match the observation to the evolutionary signal.

There is also an ethical dimension. Distributed observation can create surveillance concerns: who is being monitored, who controls the data, and who benefits? Ecological observations can reveal the locations of rare species and invite exploitation. Public health surveillance can reveal patterns that communities may not want broadcast. These are not reasons to abandon crowd-sourced science; they are reasons to treat it with the seriousness usually reserved for laboratory biosafety. Trust is an experimental condition, and it can be lost.

When citizen science works well, it creates a particular kind of intimacy with evolution. The LTEE's drama unfolds in a controlled room, measured in generations and freezer vials. Citizen science relocates that drama to sidewalks, parks, vacant lots, and household routines. It turns evolution from something that happened long ago or far away into something that leaves signatures in the places people already inhabit. The selection pressures are often human-made: salt spread on winter roads, herbicides sprayed along railway lines, artificial lights that draw nocturnal insects, buildings that create nesting opportunities for some birds and barriers for others, antibiotics and antivirals that reshape microbial populations.

There is a temptation to treat this as a feel-good story about public engagement, and it can be that. But the deeper story is about power. Humans are now one of the dominant forces shaping the selective environments of other organisms, and increasingly our own pathogens. The data needed to understand that force are distributed in space and time in the same way our influence is. A lone scientist with a clipboard cannot capture it. A crowd, equipped with simple tools and connected by shared platforms, sometimes can.

One of the quiet achievements of modern biology is that it has made ordinary acts of noticing scientifically meaningful. A photograph of a moth on a porch light can become a datum in a global map of trait variation. A sequence uploaded by a regional lab can become a branch on a worldwide tree of viral evolution. A sample from a wastewater treatment plant can become an early warning of a lineage that will, in a few months, shape hospital admissions. In each case, the underlying logic is the same as in the flask and the freezer: variation arises, selection filters it, and history leaves a record. The difference is that the record is no longer confined to a lab notebook. It is written across cities and shared across networks, in observations made by people who may never call themselves evolutionary biologists, yet who are, by the simple act of collecting and sharing evidence, helping to read evolution as it happens.

Chapter 8

Designing the Future

We often treat evolution as a chaotic, backward-looking force, yet its mechanisms are increasingly predictable. By recognizing the biological costs of adaptation and the patterns of genetic change, we can transition from passive victims of superbugs to architects of evolutionary outcomes, designing policies that steer distinct biological futures.

8.1 The Heavy Price of Survival

A useful way to think about evolution is as a series of bargains struck under pressure. A population faces a threat—an antibiotic, a pesticide, a new climate extreme—and natural selection rewards any genetic change that helps individuals survive long enough to reproduce. Yet survival is rarely a free upgrade. Most adaptations are paid for somewhere else: in energy, in speed, in accuracy, in the ability to compete when conditions return to normal. Evolution is a superb mechanic working with whatever parts happen to be lying around, not an engineer designing from scratch. When the emergency ends, yesterday's clever fix can become today's drag.

Biologists call these bargains *fitness trade-offs*. Fitness, in the strict evolutionary sense, is not strength or health or even longevity. It is reproductive success: how many descendants you leave, relative to others, in a particular environment. A mutation that improves fitness in one setting can reduce it in another. The everyday intuition is simple. You can carry a heavy winter coat through a blizzard and be grateful for every extra layer. Carry the same coat through a humid summer and it becomes a burden. Adaptations behave similarly because bodies and cells run on limited budgets. They must divide resources among growth, repair, reproduction, and defence. If more is spent on one function, less is available for another.

This is why resistance, in microbes and pests alike, so often looks impressive in the moment and disappointing in the long run. A resistant bacterium may shrug off a drug that kills its neighbours, but it might also grow more slowly, absorb nutrients less efficiently, or struggle to compete when the drug is absent. A pesticide-resistant insect may survive a sprayed field but pay for it with impaired movement or reduced fertility. The key point is not that costs *always* occur—biology is too diverse for absolutes—but that costs are common enough, and structured enough, to become an exploitable weakness.

The classic modern stage for these trade-offs is the antibiotic era. When penicillin and its successors spread through medicine in the mid-twentieth century, they changed the world so abruptly that bacteria were, in evolutionary terms, ambushed. Hospitals became landscapes saturated with lethal molecules. Sewers, farms, and households soon followed. Under such conditions, a mutation that offers even modest resistance can be the difference between extinction and explosive expansion. Natural selection, as ever, does not pause to ask whether the solution is elegant.

One of the simplest resistance mechanisms is to alter the drug's target—the cellular component the antibiotic is designed to attack. Streptomycin, for example, interferes with bacterial protein synthesis by binding to part of the ribosome. A single change in a ribosomal protein can reduce binding and confer resistance. But ribosomes are not optional accessories; they are the cell's translation machinery, tuned by billions of years of selection for speed and accuracy. Tweak them, and you often pay in slower growth or more frequent errors. In the presence of streptomycin, that price is worth paying. In a drug-free environment, it is a competitive handicap.

These costs are not mere theoretical expectations; they have been measured in straightforward experiments that resemble a sporting match. Take a resistant strain and a susceptible strain, mix them in the same flask, and let them compete for resources. In drug-free media, susceptible cells commonly win because they divide faster. Add the drug, and the ranking reverses: susceptible cells crash, resistant cells flourish. The logic is so clean that it can feel like a moral fable—hubris punished, compromise rewarded—but the reality is messier, and that messiness matters if one hopes to use trade-offs to design better strategies.

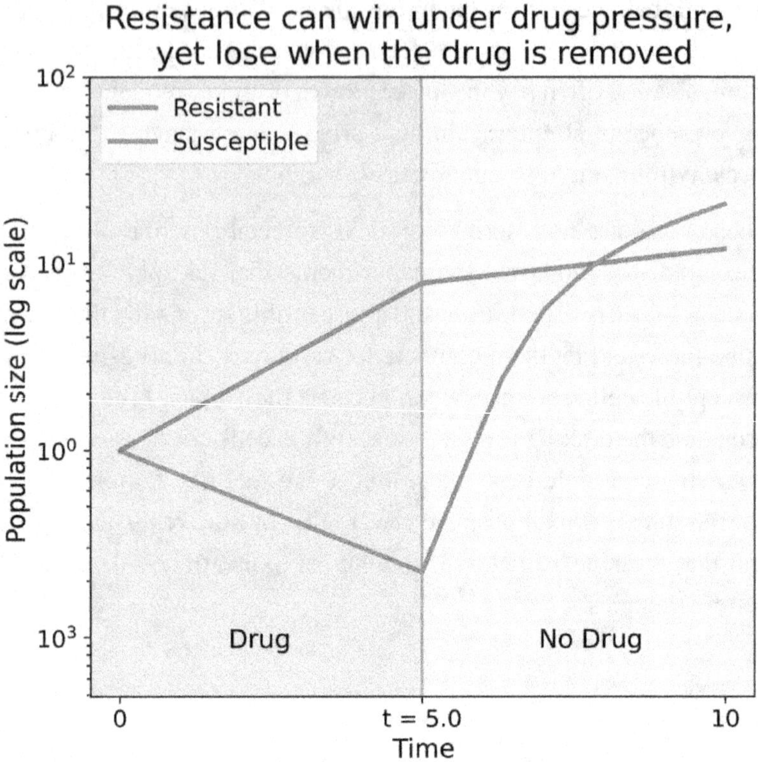

That picture—two lines crossing when the environment changes—is the intuitive heart of trade-offs. It suggests a hopeful policy: remove the antibiotic, and resistant strains should fade away, outcompeted by susceptible ones. For a time, this hope shaped real-world thinking. In some settings it works, at least partially. The catch is that evolution does not stop when the drug is withdrawn. The resistant lineage, now disadvantaged, becomes a target for selection in the opposite direction: selection for ways to keep resistance *and* regain performance.

This is where *compensatory mutations* enter the story. A compensatory mutation is a second genetic change that reduces the cost of the first. It does not necessarily remove resistance; it can simply patch up the side-effects. In the 2010s, Andersson and Hughes synthesised a large body of work showing how rapidly compensation can evolve in bacteria, sometimes within a few hundred generations—a blink in microbial time. The key lesson is that the cost of resistance is often front-loaded: it is high when resistance first appears, and lower after evolution has had time to tinker.

Consider a concrete mechanism. One common route to antibiotic resistance is to reduce the cell's permeability, making it harder for a drug to enter. Bacteria can do this by changing *porins*, protein channels in the outer membrane that normally allow nutrients to pass. The immediate benefit is obvious: fewer antibiotic molecules get inside. The cost is equally obvious: fewer nutrients get inside too. Under drug pressure, starving a little is preferable to dying. Remove the drug, and that same reduced permeability becomes a competitive liability. Yet bacteria can sometimes compensate by upregulating alternative porins or adjusting regulation so that nutrient flow improves without fully restoring vulnerability. A 2015 study of such compensation showed resistant lineages climbing back toward near-normal growth rates over roughly 250 generations, not by undoing the original mutation, but by altering the surrounding network of gene regulation. The bacterium, in effect, learned to breathe through a different set of pores.

Compensation reveals why the phrase "resistance has a cost" is both true and insufficient. The cost is real, but it is not a fixed tax. It is a moving target shaped by genetic details. Vogwill and MacLean's 2015 meta-analysis made this variability hard to ignore: fitness costs differed widely across studies and were partly explained by the genetic basis of resistance. Chromosomal resistance mutations, on average, tended to carry higher costs than plasmid-borne resistance, which helps explain why plasmids are such successful vehicles for resistance genes in nature. Plasmids—small, often circular DNA molecules separate from the bacterial chromosome—can hop between bacteria, spreading useful genes like trading cards. Carrying a plasmid can itself be costly, because it must be replicated and managed, but the burden varies: some plasmids are relatively light, some heavy, and costs often increase with the breadth of resistance they encode.

The plasmid story also highlights a second reason costs do not always rescue us: costs can be diluted by ecology. Even if a resistance plasmid burdens an individual cell, it can persist in a community if it spreads efficiently between cells, or if the environment is patchy enough that drugs are never fully absent. Modern landscapes are full of low-dose exposures: trace antibiotics in waterways, intermittent treatments in livestock, uneven drug penetration within tissues. In such settings, selection does not cleanly flip from "drug" to "no drug." Instead, it flickers, and flickering can be enough to keep resistance afloat.

The same bargain-and-compensation logic plays out beyond hospitals. Agriculture, for instance, has repeatedly witnessed the rise of resistance to pesticides and herbicides. The emotional stakes are different—crop losses instead of clinical deaths—but the evolutionary mechanics are closely related. A herbicide that blocks a plant enzyme may be defeated by a mutation in that enzyme. Yet enzymes are finely tuned machines; alter their structure and you may slow the plant's growth when the herbicide is not present. Farmers and ecologists have long looked for these penalties, hoping they might allow resistance to fade if chemical use is paused. Sometimes it happens. Sometimes resistance persists stubbornly, suggesting either low costs, rapid compensation, or continued selection in the background.

A famous early example of the cost idea, told and retold because it is so vivid, comes from the mid-century experiments with myxomatosis in Australia. The virus was introduced to control invasive rabbits, initially causing extraordinarily high mortality. Over time, the rabbits evolved resistance and the virus evolved reduced virulence; natural selection favoured rabbits that survived infection and viruses that transmitted better by not killing hosts too quickly. Although this is not antibiotic resistance, it shows the same principle: survival under intense pressure can reshape traits in ways that later look like compromises. A rabbit that invests more in immune defence may pay in growth or reproduction, and a virus that kills less aggressively may sacrifice short-term replication for long-term spread. Evolution, again, bargains.

Trade-offs are also written into the physics of molecules. Many resistance mechanisms involve pumps embedded in membranes that actively eject toxins. These *efflux pumps* are impressive: tiny machines that grab and export diverse compounds, including antibiotics. But pumping costs energy, and running pumps at high levels can divert resources from growth. Worse, broad-spectrum pumps can accidentally export useful molecules or disturb membrane balance. In a drug-free environment, a bacterium that constantly runs its pumps like bilge motors may be outcompeted by a more economical neighbour. Yet if that bacterium later acquires regulatory tweaks that keep pumps mostly off until needed, the cost shrinks. Compensation can be as simple as better management.

A further complication is that costs are not universal; they are context-dependent. A resistance mutation might slow growth in a rich laboratory medium but have little effect in a nutrient-poor environment where all cells grow slowly anyway. Or it might carry a cost at one temperature but not another. This matters because human bodies and farms are not flasks. They are heterogeneous, with gradients of oxygen, pH, nutrients, and immune pressure. A mutation can be costly in one microenvironment and beneficial in another, allowing resistant strains to persist in refuges even when drugs are withdrawn.

There is also a subtle distinction between *reversal* and *stabilisation*. One might imagine that if resistance is costly, then stopping antibiotic use should lead to a return to the ancestral, susceptible genotype. Sometimes true reversal happens: the original resistance mutation disappears. But often the path of least resistance is not reversal; it is stabilisation. Compensatory mutations can lock a lineage into a new genetic configuration that is both resistant and competitive. Once the population has climbed onto that plateau, simply removing the drug may not push it back down. Evolution is historically constrained: it depends on what variants appear and in what order. By the time the environment changes, the old road back may be blocked by new construction.

The irreversibility can be psychological as well as genetic. Public health systems, once confronted with resistant pathogens, rarely return to a world without antibiotics. They change protocols, alter prescribing, switch drugs, add combinations. The environment becomes dynamic, and bacteria adapt to that dynamism. A hospital is not a single selective pressure; it is a mosaic of them, with different drugs used in different wards, different patients receiving different courses, and different tissues within a single patient presenting different challenges. Under those conditions, a resistant genotype might never face a truly drug-free world long enough to be purged.

Even so, the cost idea remains one of the few genuinely hopeful levers available. It suggests that resistance can be *managed* rather than fantasised away. If resistance is costly, then maintaining a population of susceptible competitors can restrain resistant types by simple competition. If resistance is costly, then one can sometimes design treatment schedules that exploit those weaknesses: a drug that selects for resistance may simultaneously make bacteria more vulnerable to a second drug, an effect known as *collateral sensitivity*. The promise is not that evolution will stop, but that evolution may be steered into corners where every escape route comes with a price tag.

Yet any practical use of trade-offs requires a certain humility about what "cost" really means. Costs can be large or tiny, immediate or delayed, visible in one assay and invisible in another. They can be paid by the individual, by its offspring, or by its ecological partners. They can disappear by compensation or persist through genetic linkage. They can even be masked by interventions that change the mutation process itself: some treatments may elevate mutation rates, creating a grim paradox in which the most aggressive assault on a pathogen also increases the chance that a rare resistant mutant will appear in time to repopulate the niche. Evolutionary success is not just about surviving today; it is about keeping enough options open to survive tomorrow.

The deeper lesson is that adaptation is always entangled with constraint. When organisms evolve quickly, the speed is not proof that biology is infinitely malleable; it is proof that small genetic changes can have large effects under strong selection. Those same changes often tug on other traits because genes do not act in isolation. A mutation that alters a protein's shape may influence its stability, its interactions, its regulation, and its demand for cellular chaperones. A plasmid that brings resistance genes may also bring replication costs and conflicts with the host's control systems. A pump that exports toxins may also export signals. The heavy price of survival is paid in these hidden currencies.

Seen this way, the most valuable question is not "Does resistance have a cost?" but "What kind of cost is it, who pays it, and under what conditions does it matter?" In that question lies a kind of practical optimism. Costs are vulnerabilities. Vulnerabilities are opportunities. Evolution cannot hand out miracles without sending someone the bill.

8.2 Forecasting the Unseen

Evolution has long been narrated like detective work. A trait appears—a darker moth, a drug-resistant bacterium, a finch with a deeper beak—and biologists reconstruct the scene after the fact, inferring what pressures must have been present and what genetic changes probably occurred. That logic remains powerful, but it is also vulnerable to storytelling. With enough imagination almost any outcome can be made to sound inevitable. The more useful question, especially on human timescales, is harder and more uncomfortable: given the conditions now, what is likely to evolve next?

Forecasting evolution sounds almost like a contradiction. Mutation is random, environments wobble, and the genetic routes to similar outcomes can be many. Yet the same could be said of weather. The atmosphere is chaotic, and each gust depends on the flutter of countless molecules. Still, meteorologists forecast storms with skill that would have seemed magical a century ago, not because they eliminate uncertainty but because they identify the few variables that matter most, measure them at scale, and run models that translate physics into probabilities. Evolutionary forecasting aims for a similar shift: away from elegant after-the-fact explanations and toward forward-looking, quantitative bets.

The most familiar example arrives every autumn in the form of an influenza vaccine. The virus evolves quickly, and immunity from past infections or vaccinations pushes it to change its surface proteins in ways that help it evade antibodies. The global public-health system responds by tracking circulating strains, comparing their antigenic properties (roughly, how well antibodies recognise them), and recommending which variants should be used to manufacture the next season's vaccine. This is not fortune-telling. It is a continuous pipeline of surveillance, laboratory measurement, and modelling that turns evolutionary change into a public decision with factories, budgets, and lives at stake.

The historical anecdote embedded here is often overlooked because it is logistical rather than heroic: modern influenza forecasting is built on the slow construction of networks. After the 1918 pandemic and subsequent waves, countries began to formalise reporting and sample sharing, and by the mid-twentieth century the World Health Organization helped coordinate a global system of influenza centres and collaborating laboratories. The scientific glamour lies in the virology; the forecasting power lies in the boring infrastructure. Without consistent sampling, forecasting becomes guesswork. With sampling, it becomes an argument with data.

Even with data, influenza prediction is far from perfect. Some seasons, the chosen strains match circulating viruses well; other years, the virus takes an unexpected turn. Those misses are not just embarrassing; they are informative. They show where models are missing key inputs (such as changes in transmission patterns, population immunity, or the fitness effects of particular mutations) and where evolution itself is less repeatable. Forecasting is therefore not a single model but a discipline of continually testing predictions against reality, learning where the map does not match the territory.

Influenza is a case where the main challenge is not measuring mutations—sequencing is now routine—but translating mutations into consequences. Two viruses can differ by only a few amino acids, yet one may spread faster, escape existing immunity, or replicate better at the temperatures found in the human upper airway. Evolutionary forecasting requires a *genotype-to-phenotype map*: a way to connect genetic changes (genotype) to observable traits (phenotype) that matter for survival and reproduction. In the past, such maps were built slowly, one mutation at a time. On the timescale of pandemics and hospital outbreaks, one-at-a-time science is too slow.

A striking acceleration has come from a technique with a clunky name and a simple ambition: *deep mutational scanning*. The basic idea is to build, in the lab, a library containing a vast number of genetic variants of a particular gene or protein. One then exposes that library to a selection pressure—for a virus, perhaps antibodies in human serum; for a receptor-binding protein, perhaps the ability to attach to host cells—and uses high-throughput sequencing to see which variants become more common and which disappear. Each mutation gets an experimental "score" that approximates its effect on a trait of interest.

A helpful way to picture this is as an enormous audition. Instead of waiting for evolution to try one candidate at a time, deep mutational scanning invites thousands of candidates onto the stage at once and watches who remains standing after the spotlight and the critics do their work. The method does not bypass natural selection; it recreates a stylised version of selection under controlled conditions and at industrial scale.

During the COVID-19 pandemic, deep mutational scanning became part of the scientific response to a rapidly evolving virus. The most scrutinised viral component was the spike protein, used by SARS-CoV-2 to enter human cells and targeted by many antibodies. In 2024, a major deep mutational scanning effort mapped the effects of thousands of spike mutations in important variant backgrounds, measuring properties such as ACE2 receptor binding, cell entry, and escape from polyclonal human sera. One of the sobering findings was that mutations outside the canonical receptor-binding domain could still alter ACE2 binding and viral fitness. The intuitive focus on a single "hotspot" was too narrow: the whole protein is a mechanical and immunological system, and evolution can improve system performance through surprising parts.

This matters for forecasting because it expands the feature space. If only a small region mattered, prediction could focus on that region; if many regions matter, prediction becomes a higher-dimensional problem. Yet the same result is also encouraging. It suggests there are measurable, biophysical constraints that shape what mutations are likely to succeed. If a mutation improves immune escape but severely damages cell entry, it may never dominate in the real world. A forecast improves when it can incorporate those trade-offs.

The more philosophical point is that evolutionary forecasts often require *intermediate traits*. Natural selection acts on reproductive success, but reproductive success is the end of a long causal chain. For a virus, "fitness" depends on how well it enters cells, how fast it replicates, how effectively it transmits between hosts, and how it interacts with the immune system. For a bacterium in a patient, it depends on growth, resource acquisition, resistance mechanisms, and evasion of immune clearance. These are not all equally easy to measure, and not all equally important in every context. Forecasting is therefore partly about choosing which intermediate traits are worth measuring and which can be ignored.

There is a danger here. A laboratory assay can become a trap, offering numbers that are precise but irrelevant. An antibody-escape measurement in a dish might not translate to escape in a human population if the dominant immune pressures are different, or if the mutation changes viral transmission in ways that overwhelm its escape benefit. The discipline of forecasting demands constant calibration against reality: do the traits we measure actually predict which lineages expand?

One way evolutionists formalise these ideas is with a *fitness landscape*. Imagine each genotype as a point in a vast space, with height representing fitness. Populations tend to climb uphill by natural selection, but the landscape may be rugged, with multiple peaks separated by valleys. A mutation that is beneficial on one genetic background might be harmful on another; this interaction is called *epistasis*. Rugged landscapes make forecasting hard because evolution's next step depends on where you start and which paths are accessible without crossing fitness valleys.

Yet landscapes also provide a foothold. If one can map even a small part of the landscape—for example, the fitness of a set of mutations and their combinations—one can begin to anticipate which routes are likely and which are blocked. This is not perfect prediction; it is constraint-based prediction. In a rugged landscape, evolution cannot simply choose the globally best solution; it must move through local steps that each provide some advantage. Forecasting often rests on identifying these constraints: the set of mutations that are both available and beneficial under current conditions.

The practical use of landscapes becomes vivid in a context where humans control the selection pressures: antibiotic therapy. In a hospital, the order in which drugs are given is not an accident of weather; it is a decision. Daniel Nichol and colleagues (2015) used a Markov-chain approach to treat the evolution of resistance as transitions on a landscape, where the probability of moving from one genotype to another depends on mutation and on which drug is applied. Their work emphasised a counterintuitive property: selection is often *non-commutative*. In plain language, the order of drugs can matter as much as the drugs themselves. Giving drug A then drug B may produce a different evolutionary outcome than giving drug B then drug A, even if the total exposure is the same.

This is easy to grasp with a mundane analogy. The sequence of locks you pass through can determine where you end up, even if each lock only opens one door at a time. In genetics, each drug changes which mutations are advantageous, shaping which genotypes are reachable next. A therapy sequence can accidentally guide a population toward a highly resistant genotype, the way a poorly designed series of detours can funnel traffic into a jam. The unsettling implication is that naive cycling—switching drugs periodically in hopes of "staying ahead"—can backfire if the cycle happens to follow an evolutionary corridor.

Here, forecasting becomes a kind of applied ethics. The patient needs to be treated now, not in an idealised future with perfect models. The aim of forecasting is not to delay therapy but to choose regimens that reduce the probability of failure. Even a modest predictive advantage can matter when the downside is a resistant infection that is harder to treat, more expensive, and more likely to spread.

A particularly tantalising forecasting concept is *collateral sensitivity*: the idea that evolution toward resistance to one drug can make an organism more sensitive to another. This can occur for mechanistic reasons. For example, if resistance is achieved by overexpressing an efflux pump that exports drug A, the pump might also alter membrane properties or cellular energy balance in a way that increases susceptibility to drug B. Alternatively, a mutation that changes a target protein to evade one antibiotic might inadvertently expose a weakness exploited by another. If these patterns are reliable, they can be used to design drug sequences that steer evolution into a trap.

The optimism is tempered by a crucial caveat: repeatability. A 2018 study in *Nature Communications* underscored that collateral sensitivity patterns can depend on which evolutionary path was taken, even under similar selection. Two replicate populations exposed to the same drug can evolve different resistance mechanisms, and those mechanisms can have different collateral effects. This is contingency: history matters. A forecasting model that assumes a single, clean collateral sensitivity pattern may be brittle in the face of real biological variability.

This is where evolutionary forecasting begins to resemble risk management more than prophecy. The goal is not a single prediction of what *will* happen, but a distribution of what *could* happen, with estimates of likelihood. In practice, this might mean choosing a therapy sequence that performs reasonably well across many plausible evolutionary routes rather than spectacularly well in one idealised route. Robustness, in this context, is a virtue.

Forecasting also forces a shift in how one treats "randomness." Mutation is random with respect to what an organism needs, but it is not random in every sense. Mutation rates vary across genomes; some changes are more likely than others; some require a single nucleotide substitution while others require multiple coordinated changes. And natural selection is not random at all: it is a systematic filter. A forecast improves when it incorporates the unevenness of mutation supply and the unevenness of selection.

The pandemic offered a public demonstration of this logic. When particular spike mutations began to appear independently in different lineages, it suggested that those changes were both accessible and beneficial under prevailing conditions. Convergent evolution—the repeated appearance of similar solutions—is one of the strongest hints that forecasting is possible. If evolution keeps reaching for the same tools, then the toolbox is constrained. Deep mutational scanning, combined with surveillance, can make those constraints legible before a variant becomes dominant, allowing public-health responses to move from reactive to anticipatory.

There is also an emotional dimension to forecasting that is easy to miss when discussing models. Retrospective stories can be comforting: they imply that events unfolded for coherent reasons, that the world makes sense after the fact. Forecasts, by contrast, are unsettling because they are probabilistic. They ask us to accept uncertainty and still act. A forecast that says "there is a 30% chance of a high-resistance lineage emerging under this regimen" is not a satisfying narrative. It is, however, the kind of statement that can guide policy if we learn to treat probabilities as actionable information rather than as admissions of ignorance.

Perhaps the most honest promise of evolutionary forecasting is this: it can turn surprise into a measurable quantity. Instead of being shocked that resistance emerged, one can ask whether resistance emerged faster than expected, by a different route than expected, or in a context where it was supposed to be unlikely. Each surprise becomes data for improving the model. In a world where evolution happens on hospital timescales and agricultural seasons, that feedback loop—predict, observe, revise—is one of the few ways to keep pace with change.

What makes the enterprise plausible is not a belief that evolution is simple, but a recognition that evolution is constrained by chemistry, by genetics, and by history. Mutations must still make proteins fold. Viruses must still enter cells. Bacteria must still balance membranes, energy, and growth. Landscapes may be rugged, but they are not arbitrary. The unseen future is not a blank page; it is a limited set of options filtered by the realities of living matter, with selection doing the accounting in real time.

8.3 The Architecture of Resilience

There is a stubborn fantasy that shows up whenever humans fight evolving enemies: the fantasy of the clean win. One imagines a hospital so antiseptic that microbes have nowhere to hide, or a farm so thoroughly treated that pests simply vanish. The fantasy feels modern and decisive, an engineer's solution to a biological problem. Yet in systems that can evolve, total victory is often the fastest route to a new defeat. The more completely a drug or pesticide wipes out the susceptible majority, the more it hands the future to whatever rare survivors remain.

Resilience offers a different mindset. In ecology, C. S. Holling's work in 1973 made a sharp distinction between *stability* and *resilience*. A stable system resists change and returns quickly to an equilibrium after a small disturbance. A resilient system can absorb larger shocks without flipping into a fundamentally different state. The difference matters because many real systems have multiple "basins" of behaviour: clear lakes can become turbid, diverse grasslands can become shrublands, and a hospital's microbial community can become dominated by a handful of difficult pathogens. The goal is often not to freeze a system in place—that is rarely possible—but to keep it within a regime where it remains functional and recoverable.

Biomedicine and agriculture have begun to borrow this ecological idea and translate it into design. The result is sometimes called *evolution-proofing*: strategies that do not assume evolution can be halted, but aim to shape it, slow it, and channel it into less harmful forms. It is not a single technique. It is an architecture: a way of arranging pressures, refuges, redundancies, and feedback so that an evolving population remains manageable.

To understand why architecture matters, it helps to return to the basic mechanics of natural selection. Selection needs variation. Variation can come from mutation, from gene transfer, or from hidden diversity already present. When the environment becomes harsh, selection amplifies whatever variants can cope. If the environment is harsh everywhere, all the time, selection becomes both strong and simple: there is one obvious target for adaptation. In contrast, if the environment varies across space and time—patches with drug, patches without; days with pesticide, days without; tissues where a drug penetrates well and tissues where it does not—then there are multiple ways to succeed, and no single genotype dominates as easily.

That variability is often treated as a nuisance, an imperfection to be eliminated. But it can be turned into a tool. A key insight from resistance management is that maintaining a population of susceptible organisms can be protective. Susceptible competitors can dilute resistance genes, outcompete resistant types when the pressure is absent, and reduce the chance that resistance becomes fixed—that is, present in essentially everyone.

The archetypal agricultural version of this idea is the *refuge strategy*. It became widely discussed with the adoption of genetically engineered crops producing *Bt* toxins (from *Bacillus thuringiensis*), which are lethal to certain insect pests. The premise of Bt crops is elegant: make the plant produce a protein that targets pests, reducing the need for sprayed insecticides. The evolutionary problem arrives just as elegantly: if the toxin is everywhere, then any insect with a resistance allele is strongly favoured, and resistance can spread quickly.

Refuges are an attempt to build a counterweight into the system. Farmers set aside areas planted with non-Bt crops (or otherwise untreated), where susceptible insects can survive. Those susceptible insects then mate with any resistant survivors emerging from the Bt fields. If resistance is not fully dominant—that is, if heterozygous offspring are still vulnerable to the toxin—then the refuge can slow the rise of resistance by ensuring that resistant insects rarely encounter each other and produce fully resistant offspring. The refuge does not eliminate pests; it manages their evolution.

The logic can sound, at first, like heresy. Why would one intentionally allow pests to live? Because the alternative is often worse: a rapid transition to a pest population that the toxin can no longer control, forcing a return to broader-spectrum insecticides and the ecological collateral damage that comes with them. The refuge is not kindness. It is a design feature that treats susceptibility as a resource worth conserving.

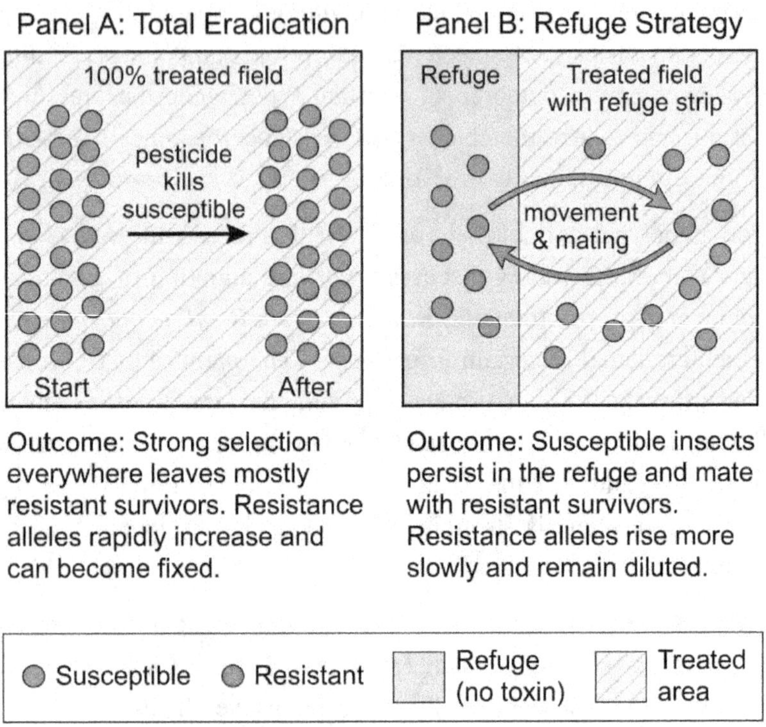

The diagram captures a trade that is easy to miss in arguments about "control" and "eradication." Total treatment feels like maximal control, but it simplifies selection so much that it accelerates adaptation. A refuge feels like tolerating failure, but it complicates selection enough to slow adaptation and preserve options.

The same architecture can be applied in medicine, though the word "refuge" sounds strange when the "pest" is a pathogen in a patient's body. Here, the practical constraints are different. A farmer can plant a strip of non-Bt crop; a clinician cannot ethically set aside a strip of untreated infection. Even so, the underlying design principle can still appear: avoid creating a world in which only the most resistant organisms can survive.

One medical analogue is the debate over aggressive, maximal-dose therapy versus strategies that aim to maintain competition. Some cancers, for example, contain drug-sensitive and drug-resistant cells. If treatment wipes out the sensitive cells completely, it can remove the competition that kept resistant cells in check, allowing the resistant population to expand rapidly. In that context, adaptive therapy has been proposed: doses are adjusted to control tumour burden while preserving enough sensitive cells to suppress resistant ones. The idea is not to surrender to the disease but to manipulate the tumour's ecology. The evolutionary opponent is not a single clone but a shifting community, and community management can sometimes outperform brute force.

Antibiotics raise similar questions, though the analogy must be handled carefully because infections can become life-threatening quickly. Still, the same dynamics can occur in miniature. A patient may harbour a mixed population of bacteria, some more susceptible, some less. A drug that penetrates unevenly can create spatial refuges inside the body: pockets where bacteria experience sublethal concentrations. Those pockets are not designed; they are accidents of physiology. But once acknowledged, they can influence how regimens are constructed and how long they are maintained.

The deeper point is architectural: selection pressures are not just about *strength* but about *shape*. A pressure can be strong and uniform, or strong and patchy, or mild but persistent, or intermittent. Each shape creates different evolutionary incentives. In engineering, a bridge is not made safe only by adding more steel; it is made safe by placing supports where stress concentrates and by designing redundancy so that a single failure does not bring the whole structure down. Evolution-proofing works similarly: one designs the environment so that the "stress" of selection does not funnel the system into a single catastrophic outcome.

Biological robustness theory helps explain why certain architectures work. Hiroaki Kitano's 2004 synthesis emphasised that robustness in living systems often arises from feedback control, redundancy, modularity, and distributed function. These features allow a system to keep working when parts fail. They also produce a familiar trade: robustness in one dimension can create fragility in another. A pathogen robust to one antibiotic mechanism may become fragile to another; a pest robust to one toxin may pay a cost elsewhere. The trick is to design interventions that *create* these trade-offs rather than accidentally erasing them.

Feedback is particularly important. In many resistance crises, the problem is not simply that a drug selects for resistance, but that the system continues to apply the same selection even as resistance rises. Feedback would mean adjusting behaviour based on signals: prescribing patterns that respond to local resistance data, infection control that adapts to outbreaks, agricultural practices that change as resistance markers become common. Feedback requires measurement and the willingness to act on it. Without it, management is blind.

Modularity is another architectural concept with practical implications. When one antibiotic is used for everything, the selection pressure is global. When therapies are tailored and diversified, pressures become more modular, reducing the chance that a single resistance mechanism becomes universally advantageous. The same logic applies to crops and pesticides. Monocultures, whether genetic or chemical, simplify selection. Diversity complicates it. Diversity is not a sentimental preference for variety; it is a mechanism for preventing any one adaptive solution from sweeping through a system.

A third architectural idea is *neutral space*, emphasised by Andreas Wagner (2005) in discussions of robustness and evolvability. In genetics, a neutral space is a set of genotypes that differ from one another but have similar fitness under current conditions. Neutral spaces matter because they allow populations to drift without immediate penalty, accumulating hidden variation that can become useful when conditions change. This can sound dangerous: it suggests that robustness creates the raw material for future adaptation. But it also suggests an opportunity. If one can keep an evolving population in a broad neutral space where many genotypes are acceptable, one may avoid pushing it onto a narrow peak where resistance is locked in and hard to dislodge. Control, paradoxically, can come from allowing multiple workable configurations rather than insisting on a single, maximally optimised one.

In agricultural resistance management, the refuge strategy is often paired with the "high-dose" idea: apply a sufficiently high toxin dose in treated areas to kill not only susceptible pests but also those with partial resistance, while the refuge sustains susceptibility. The architecture is deliberate: strong selection in one region, no selection in another, with gene flow between them. It is a spatial feedback system built from fields and insect movement. When it works, it works because it manipulates mating patterns and genetic dominance. When it fails, it often fails for architectural reasons too: refuges too small, poor compliance, pests with high mobility, resistance that is dominant, or ecological factors that reduce mixing.

There is a quiet ethical lesson in this. Evolution-proofing is rarely a matter of discovering a single perfect molecule. It is about governance: incentives for farmers to plant refuges, incentives for clinicians to steward antibiotics, surveillance systems that detect change early, and social trust that allows coordinated action. Architecture is not only biological; it is institutional.

The human emotional obstacle is impatience. When crops are threatened or a loved one is sick, the urge is to use the strongest tool immediately and everywhere. Sometimes that is exactly the right choice. But as a default philosophy for interacting with evolvable systems, maximal pressure can be self-defeating. Resilience thinking asks for a more mature stance: not passive, not indulgent, but strategic over time.

A resilient architecture accepts that some level of the adversary will remain. It makes room for susceptibility and diversity because these are the raw materials of control. It uses variation in space and time to prevent evolutionary monopolies. It treats selection pressures as levers that can be arranged, not merely intensified. And it measures outcomes continuously, because in a living system the target is always moving.

In the end, the most reliable victory over evolution is not to defeat it, but to recruit it: to design environments where the easiest evolutionary responses are the ones we can live with, and where the costs of survival fall hardest on the traits that threaten us most.

www.ingramcontent.com/pod-product-compliance
Lightning Source LLC
LaVergne TN
LVHW022212200926
848481LV00006B/1935